T0224068

Communications in Computer and Information Science   **660**

Commenced Publication in 2007
Founding and Former Series Editors:
Alfredo Cuzzocrea, Dominik Ślęzak, and Xiaokang Yang

More information about this series at http://www.springer.com/series/7899

Jian Chen · Yoshiteru Nakamori
Wuyi Yue · Xijin Tang (Eds.)

Knowledge and Systems Sciences

17th International Symposium, KSS 2016
Kobe, Japan, November 4–6, 2016
Proceedings

 Springer

Editors
Jian Chen
Tsinghua University
Beijing
China

Yoshiteru Nakamori
School of Knowledge Science
JAIST
Nomi, Ishikawa
Japan

Wuyi Yue
Department of Information Science
  Faculty of Science and Engineering
Konan University
Higashinada-ku, Kobe
Japan

Xijin Tang
Academy of Mathematics and Systems
  Science
Chinese Academy of Sciences
Beijing
China

ISSN 1865-0929   ISSN 1865-0937 (electronic)
Communications in Computer and Information Science
ISBN 978-981-10-2856-4   ISBN 978-981-10-2857-1 (eBook)
DOI 10.1007/978-981-10-2857-1

Library of Congress Control Number: 2016954198

Printed on acid-free paper

This Springer imprint is published by Springer Nature
The registered company is Springer Nature Singapore Pte Ltd.
The registered company address is: 152 Beach Road, #22-06/08 Gateway East, Singapore 189721, Singapore

Preface

The annual International Symposium on Knowledge and Systems Sciences aims to promote the exchange and interaction of knowledge across disciplines and borders so as to explore new territories and new frontiers. With over 16 years of continuous endeavors, attempts to strictly define knowledge science may still be ambitious, but a very tolerant, broad-based, and open-minded approach to the discipline can be taken. Knowledge science and systems science can complement and benefit each other methodologically.

The First International Symposium on Knowledge and Systems Sciences (KSS 2000) was initiated and organized by the Japan Advanced Institute of Science and Technology (JAIST) in September of 2000. Since then KSS 2001 (Dalian), KSS 2002 (Shanghai), KSS 2003 (Guangzhou), KSS 2004 (JAIST), KSS 2005 (Vienna), KSS 2006 (Beijing), KSS 2007 (JAIST), KSS 2008 (Guangzhou), KSS 2009 (Hong Kong), KSS 2010 (Xi'an), KSS 2011 (Hull), KSS 2012 (JAIST), KSS 2013 (Ningbo), KSS 2014 (Sapporo), and KSS 2016 (Xi'an) have been held successfully, with contributions by many scientists and researchers from different countries. During the past 16 years, people interested in knowledge and systems sciences have formed a community, and an international academic society has existed for 13 years.

This year KSS will be held in during November 4–6, 2016, Kobe, Japan, to provide opportunities for presenting interesting new research results, facilitating interdisciplinary discussions, and leading to knowledge transfer under the theme of "Systems Approaches to Knowledge, Technology, and Service Management". We are particularly fortunate to have three distinguished scholars to deliver the keynote speeches, which reflect the diverse features of KSS topics,

- Takayuki Ito (Nagoya Institute of Technology, Japan), "Crowd Decision-Making and Consensus Support System Based on Agent and AI Technologies"
- Tetsuo Sawaragi (Kyoto University, Japan), "Semiotic Analysis of Behavioral Knowledge and Skills for Cyber-Physical Co-Coaching"
- Alfred Taudes (Vienna University of Economics, Austria), "Production as a Service"

The organizers of KSS 2016 received 48 submissions, and finally 21 submissions were selected for the proceedings after a rigorous review process. The co-chairs of the international Program Committee made the final decision for each submission based on the reports from the reviewers, who came from Australia, China, France, Japan, New Zealand, and the USA.

To make this event possible, we received a lot support and help from many people and organizations. We would like to express our sincere thanks to the authors for their remarkable contributions, all the Technical Program Committee members for their time and expertise in reviewing the papers within a very tight schedule, and the publisher Springer for their professional help. This is the first time that KSS proceedings are published as a CCIS volume by Springer. We greatly appreciate our three distinguished scholars for accepting our invitation to deliver keynote speeches to the symposium. Last but not least, we are indebted to the local organizers for their hard work for the symposium.

The conference facilitated not only cross-cultural learning and integration, but also academic achievements in the area of knowledge and systems sciences.

September 2016   Jian Chen
                                                              Yoshiteru Nakamori
                                                                    Wuyi Yue
                                                                    Xijin Tang

Organization

KSS 2016 is organized by the International Society for Knowledge and Systems Sciences and hosted by Konan University in Japan.

General Chairs

Jian Chen	Tsinghua University, China
Yoshiteru Nakamori	Japan Advanced Institute of Science and Technology, Japan
Wuyi Yue	Konan University, Japan

Program Committee Chairs

Van-Nam Huynh	Japan Advanced Institute of Science and Technology, Japan
Tieju Ma	East China University of Science and Technology, China
Xijin Tang	CAS Academy of Mathematics and Systems Science, China
Jiangning Wu	Dalian University of Technology, China

Technical Program Committee

Quan Bai	Auckland University of Technology, New Zealand
Meng Cai	Xidian University, China and Harvard University, USA
Jindong Chen	China Academy of Aerospace Systems Science and Engineering, China
Zengru Di	Beijing Normal University, China
Serge Galam	Sciences Po and CNRS, France
Van-Nam Huynh	Japan Advanced Institute of Science and Technology, Japan
Cheng-Siong Lee	Monash University, Australia
Xianneng Li	Dalian University of Technology, China
Yongli Li	Northeastern University, China
Zhenpeng Li	Dali University, China
Bo Liu	CAS Academy of Mathematics and Systems Science, China
William Liu	Auckland University of Technology, New Zealand
Yijun Liu	CAS Institute of Policy and Management, China
Tieju Ma	East China University of Science and Technology, China
Mina Ryoke	University of Tsukuba, Japan
Xijin Tang	CAS Academy of Mathematics and Systems Science, China
Jing Tian	Wuhan University of Technology, China
Haibo Wang	Texas A&M International University, USA
Mingzheng Wang	Dalian University of Technology, China
Cuiping Wei	Yangzhou University, China
Jiangning Wu	Dalian University of Technology, China
Haoxiang Xia	Dalian University of Technology, China
Hongbin Yan	East China University of Science and Technology, China

Abstracts of Keynotes

Crowd Decision-Making and Consensus Support System Based on Agent and AI Technologies

Takayuki Ito

Department of Computer Science and Engineering,
Graduate School of Engineering, Nagoya Institute of Technology,
Gokiso, Showa, Nagoya 466–8555, Japan
ito.takayuki@nitech.ac.jp

Abstract. Much attention has been focused on the collective intelligence of people worldwide. Interest continues to increase in online democratic discussions, which might become one of the next generation methods for open and public forums. To harness collective intelligence, incentives for participants are one critical factor. If we can incentivize participants to engage in stimulating and active discussions, the entire discussion will head in fruitful ways and avoid negative behaviors that encourage "flaming." "Flaming" means a hostile and insulting interaction by Wikipedia. In our work, we developed an open web-based forum system called COLLAGREE that has facilitator support functions and deployed it for an internet-based town meeting in Nagoya as a city project for an actual town meeting of the Nagoya Next Generation Total City Planning for 2014–2018. Our experiment ran on the COLLAGREE system during a two- week period with nine expert facilitators from the Facilitators Association of Japan. The participants discussed four categories about their views of an ideal city. COLLAGREE registered 266 participants from whom it gathered 1,151 opinions, 3,072 visits, and 18,466 views. The total of 1,151 opinions greatly exceeded the 463 opinions obtained by previous real-world town meetings. We clarified the importance of a COLLAGREE-type internet based town meeting and a facilitator role, which is one mechanism that can manage inflammatory language and encourage positive discussions. While facilitators, who are one element of a hierarchical management, can be seen as a top-down approach to produce collective discussions, incentive can be seen as a bottom-up approach. In this talk, we also focus on incentives for participants and employ both incentives and facilitators to harness collective intelligence. I propose an incentive mechanism for large-scale collective discussions, where the discussion activities of each participant are rewarded based on their effectiveness. With these incentives, we encourage both the active and passive actions of participants. In this talk, I will present current results about this project.

Semiotic Analysis of Behavioral Knowledge and Skills for Cyber-Physical Co-Coaching

Tetsuo Sawaragi

Department of Mechanical Engineering and Science,
Graduate School of Engineering, Kyoto University's, Kyoto, Japan

Abstract. Recently robotic techniques are applied not only to the manufacturing fields but also to everyday life domain (i.e., ambient intelligence). Wherein, human beings are observed by a variety of sensors existing in the environment, and their invisible internal states like intentional and emotional states are inferred based on the observed sensor data so that the system can understand their activities. In this talk, we regard understanding human bodily motions and/or gestures are "semiotic processes to structuralize the infinite into some coherent internal constructs within the observer." Like a verbal language, human bodily motions are organized along the two different sorts of contexts; syntagmatic and paradigmatic contexts. Moreover, performing bodily motions is exactly an activity of communication between the actor and the cognizer (i.e., observer), where the two separate semiotic processes within the actor and the observer are interchanging with each other, and thus the meanings come out and become shared between them. As for the semiosis of human bodily motions, we approach to the constructive semiosis from two different but interconnected perspectives; from an observer's view and from an actor's view. The former deals with how a cognizer understands an actor's bodily motion emitting a plenty of "signs", while the latter is how to design actions to be understood by an observer. Our research group in the project is developing techniques for exploring the systematic structures implied in the human bodily motions. In order to extract the semiotic structures organized via syntagmatic and paradigmatic contexts, we developed two methodologies by extending the conventional methods of singular spectrum transformation (SST) and singular value decomposition (SVD), respectively. Based on the results of the above analysis, we show a testbed of cyber-physical co-coaching system, where a novice sport player is coached by being provided with semiotically analyzed behaviors and constructed 'pantomimed' motions in display as well as with other verbal instructions by an expert. It is shown how the skills of the players evolve as they accumulate such interactions, and the interface design issues for cyber-physical co-coaching system is discussed. Then, we introduce a schema theory to explain how we learn and perform discrete perceptual motor skills. Discrete motor skills are skills that take a short time to perform and involve using our senses to understand what is happening and then using our bodies to take action. The computational model of how the schema is learned, acquired and reconstructed through the iteration of assimilation and accommodation from the trials is presented for a simple target-tracking task. Finally, general issues for human-system collaborative system design and how the novel knowledge is expected to emerge out of their interactions.

Production as a Service

Alfred Taudes

Institute of Production Management, Vienna University of Economics,
Vienna, Austria

Abstract. The Internet of People has transformed the world. On the horizon is a
new Internet of Things, where things start to sense their environment, communicate via the Internet and behave autonomously as cyber-physical systems.
In Industry 4.0 machines can sense their condition and communicate with parts,
and consumer products offer new functions and generate Big Data as the basis of
novel information-based services. 3D printing and blockchain technology offer
totally new ways of production and coordination. This new wave of innovation
poses significant opportunities and threats to traditional manufacturing companies. They not only have to acquire new technical skills related to software and
data analytics but also have to adapt their business model to make full use of the
new possibilities. Analyzing the various technological options it turns out that
they have to learn and adopt business practices that were hitherto only used by
service companies and information businesses. They also have to define their
role in a novel organization of manufacturing. Real-world cases are used to
illustrate the new concept of Production as a Service.

Contents

Team Knowledge Formation and Evolution Based on Computational Experiment

Yutong Li[1(✉)] and Yanzhong Dang[2(✉)]

[1] School of Psychology, Liaoning Normal University, Dalian, China
dearliyutong@163.com
[2] Institute of System Engineering, Dalian University of Technology, Dalian, China
yzhdang@dlut.edu.cn

Abstract. In knowledge intensive team, team knowledge referred to team-level knowledge emerged from knowledge interaction among members, is vital resources for enterprise innovation. So how the team knowledge forms and evolves is what this paper concerns. Firstly knowledge interaction process is described according to member's psychological and behavioral activities, then team knowledge emerging process in knowledge interaction is depicted based on members' memories, after that a task driven-artificial knowledge intensive team is established with computational experiment method by simulating knowledge interactions to achieve team knowledge formation and evolution. According to the experiments, the influences of team scale, team knowledge space, member's knowledge learning ability, knowledge interaction willingness and initial knowledge state on team knowledge formation and evolution are analyzed. The experiments results can provide reliable decision supports for managers to use team knowledge to improve enterprise innovation.

Keywords: Team knowledge · Formation and evolution · Knowledge intensive team · Computational experiment

1 Introduction

Knowledge intensive team, as an important organizational form of innovation, has been paid more attention by enterprises [1]. Knowledge is the indispensable resources in knowledge intensive team. It has not only member-level knowledge like individual knowledge, but also team-level knowledge such as team knowledge. Team knowledge is not a simple-summation of individual knowledge; instead it is generated by knowledge interaction among members, which can coordinate members' behavior and promote members' cooperation, is vital resources to promote enterprise innovation [2,3]. Accordingly, enterprises are increasingly focusing on knowledge intensive team building, hoping to enhance its own innovation capabilities by promoting team knowledge formation and development.

From the existing achievements, the research on team knowledge mainly depends on the team-level cognitive activities [4]. Team knowledge refers to

© Springer Nature Singapore Pte Ltd. 2016
J. Chen et al. (Eds.): KSS 2016, CCIS 660, pp. 1–14, 2016.
DOI: 10.1007/078-981-10-2857-1_1

an emergent structure in which knowledge that is critical to team functioning is organized, represented and distributed within a team [5]. It has two important representations respectively shared mental models and transactive memory systems [6]. Shared mental models emphasize the sharing of individual knowledge [7]. It means that members have a common understanding of team environment; this common understanding is conducive to predict members' needs and behaviors and then improves team performance [8,9]. Transactive memory systems emphasize the distribution of individual knowledge [10]. Members clearly know "what kind of expertise that other members have"; this common cognition effectively promotes members' cooperation [11] and increases team productivity [12,13]. Therefore, team knowledge as an emergent state, is team-level knowledge generated by knowledge interaction among members. It is the unification of homogeneous knowledge and heterogeneous knowledge. And it's beneficial to promote team performance. So how the team knowledge emerges and evolves is what this paper concerns.

From the system perspective, knowledge intensive team is a complex system. Member as system elements have psychological activities including memory, decision, evaluation, etc., and behavioral activities, knowledge interaction, executing tasks and so on. In behavioral activities, knowledge interaction is a complex process where many activities are interrelated such as contact, understanding, communication and evaluation. It is team knowledge that is system characteristic emerged from knowledge interaction and is derived from the bottom-up process of individual cognition. Due to the complexities of member and knowledge interaction, team knowledge emerging process is extremely complex that is difficult to quantitative representation and to use mathematical model describing its dynamic evolution. So computational experiment method is used to overcome the above difficulties in this paper. Computational experiment [14] aids computers to construct experimental object and environment, simulates agents' adaptive interaction based on bottom-up agent modeling approach to explain system's emergence mechanisms and evolvement laws. It makes up for the shortcomings of traditional qualitative and quantitative research methods, and as a new social science research method has attracted many scholars to join this field research [15–17]. Thus we adopt computational experiment to construct a task-driven artificial knowledge intensive team by simulating knowledge interaction among members to achieve and study team knowledge formation and evolution.

2　Description of Team Knowledge Formation and Evolution

Knowledge intensive team is a task-driven team. Each member having psychological and behavioral activities completes the assigned tasks actively. Figure 1 describes a member's psychological and behavioral activities when he undertakes a task. For a member, task is external stimulus; knowledge need is the gap between knowledge required for task and his individual knowledge. When knowledge need is less than or equal to zero, member can fulfill this task and

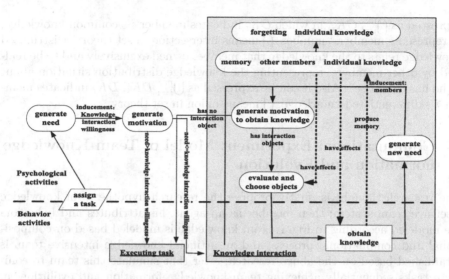

Fig. 1. A member's psychological and behavioral activities when he undertakes a task

produce motivation of performing task. Conversely, when it is greater than zero, member can't fulfill task and produce motivation of acquiring knowledge via knowledge interaction. Knowledge interaction is a complex process. It involves not only member's psychological activities, such as memory, forgetting, decision-making and evaluation, but also behavioral activities, like knowledge interaction and executing tasks, etc. In psychological activities, memory is formed based on memorizing other members' individual knowledge during knowledge interaction; in turn, memory also helps member to evaluate and select interaction objects. At the same time, memory exists the phenomenon of forgetting, mainly refers to forgetting other members' individual knowledge. So team knowledge is emerged gradually from the knowledge interaction process above, and is mainly derived from members' memories.

According to the relationship between knowledge and members, each member's memory contents can be divided into two categories, one is the knowledge which he has and other members have, called common knowledge; the other is the knowledge which he doesn't have but other members have, called differentiated knowledge. So the two sides in knowledge interaction gradually form the cognitive memory of what common knowledge they both have and of what differentiated knowledge that other side has, and the knowledge intensive team forms team knowledge gradually with several knowledge interactions. Team knowledge is the shared understandings of "what kind of common knowledge all members have" and of "who has what kind of differentiated knowledge". This above process is cross-level emergence phenomenon. According to "compositional emergence" and "compilational emergence" [18], all the members' common knowledge is composed of shared knowledge, and all the members' differentiated knowledge is compiled of distributed knowledge. Shared knowledge refers to the knowledge that all members have, representing the knowledge sharing situation among members. It can be expressed as $\bigcap_{i=1}^{m} CK_i$, in which CK_i indicates member i's common knowledge, m represents members' number, $\bigcap$ means intersection in set theory. Distributed knowledge refers to the knowledge that member owned exclusively and to be realized by other members, representing the knowledge distribution situation about "who has expertise". Also it can be expressed as $\bigcup_{i=1}^{m} DK_i$, DK_i indicates member i's differentiated knowledge, $\bigcup$ means union in set theory.

3   Computational Experiment Model of Team Knowledge Formation and Evolution

The research thought is: firstly propose the research hypotheses of knowledge intensive team system; then member as an agent, his attributes and behaviors are modeled according to Fig. 1, team knowledge is modeled based on compositional and compilational process, and an artificial knowledge intensive team is established based on the above research; after that simulate this team to complete tasks sequentially achieving team knowledge formation and evolution, in which the above model is called computational experiment model of team knowledge formation and evolution; finally run this model to analyze and observe team knowledge formation and evolution.

The research hypotheses are as follows. (1) Knowledge intensive team is a task-driven team and it is heterogeneous that composed members have different knowledge. (2) Members are regard as agents having the abilities of memory, learning, decision making and tasks performing. (3) Team knowledge emerged from knowledge interaction among members reflects system characteristic.

3.1   Task Modeling

Let $P = \{p_1, p_2, \ldots p_m\}$ denote a finite set of members, $p_i \in P, i = 1, 2 \ldots m$. Let $K = \{k_1, k_2, \ldots k_n\}$ denote a finite set of knowledge, $k_c \in K, c = 1, 2 \ldots n$. Tasks number is U. $Task = \{tk_1, tk_2, \ldots kt_U\}$ denotes a finite set of tasks. For $tk_u \in Task, u = 1, 2 \ldots U$, each task has u' sub-tasks as $tk_u = \{tk_{u1}, tk_{u2}, \ldots tk_{uv}\}$, $v \in [1, u']$. tk_{uv} is described as:

$$K_{tk_{uv}} = [k_{tk_{uv},1}, k_{tk_{uv},2}, \ldots, k_{tk_{uv},n}] \tag{1}$$

$k_{tk_{uv},c} \in [0.5, 1]$, $c = 1, 2 \ldots n$, represents the amount of c kind knowledge needed in tk_{uv}.

3.2   Member Modeling

Attributes Modeling. Member as agent has knowledge attribute, behavioral attribute and psychological attribute. Knowledge attribute includes knowledge which is the resources for members performing tasks. Behavioral attribute contains knowledge learning ability. Psychological attribute includes knowledge interaction willingness, knowledge forgetting ability and memory. Knowledge interaction willingness is the intention member prefers to knowledge interaction. Memory refers to member memorizes other members' knowledge during knowledge interaction. And Knowledge forgetting ability reflects memory's forgetting phenomenon. Based on the analysis,

(1) p_i's knowledge attribute represents as:

$$K_i = [k_{i,1}, k_{i,2}, \ldots, k_{i,n}] \tag{2}$$

$0 \leqslant k_{i,c} \leqslant 1, c = 1, 2 \ldots n$. $k_{i,c}$ is the amount of c kind knowledge that p_i has.

(2) p_i's behavioral attribute represents as $a_i \in [0, 1]$.

(3) In p_i's psychological attribute, knowledge interaction willingness as $b_i \in [0, 1]$, knowledge forgetting ability as $c_i \in [0, 0.5]$ and memory described as:

$$Memory\_K_i = (u_{ijr})_{m \times n} = \begin{pmatrix} u_{i11} & u_{i12} & \cdots u_{i1n} \\ u_{i21} & u_{i22} & \cdots u_{i2n} \\ \vdots & \vdots & \ddots \vdots \\ u_{im1} & u_{im2} & \cdots u_{imn} \end{pmatrix} \tag{3}$$

$u_{ijr} \in [0, 1]$, it means p_i knows "p_j has r kind knowledge", and the amount of r p_j has is u_{ijr}.

According to $Memory\_K_i$, $Memory\_CK_i = (ck_{ijr})_{m \times n}$ indicates p_i's common knowledge, $Memory\_DK_i = (dk_{ijr})_{m \times n}$ indicates p_i's differentiated knowledge. In knowledge intensive team, if a member has a certain kind of knowledge in a low amount, this kind knowledge won't make any sense in knowledge interaction. So we assume $p_i \in P$, $k_{i,c} \in K_i$, $\varepsilon_1 \in [0, 1]$ as threshold, then

$$ck_{ijr} = \begin{cases} 0 & u_{iir} < \varepsilon_1 \\ min\{u_{ijr}, u_{iir}\} & u_{iir} \geqslant \varepsilon_1 \end{cases} \tag{4}$$

ck_{ijr} means p_i knows "he and p_j has r kind knowledge", and the amount of r they shared is ck_{ijr}.

$$dk_{ijr} = \begin{cases} u_{ijr} & u_{iir} < \varepsilon_1, \frac{u_{ijr}}{u_{iir}} \geqslant \varepsilon_2 \\ 0 & u_{iir} \geqslant \varepsilon_1 \end{cases} \tag{5}$$

When p_i has less amount of r kind knowledge than ε_1, and he knows that p_j has amount of r is more than ε_2 times of its own, k_r is defined as p_j's expertise. $\varepsilon_2 \in N^+$ is a threshold. dk_{ijr} means p_i knows "p_j has r kind knowledge as expertise", and the amount of r p_j has is dk_{ijr}.

Behaviors Modeling. Shown in Fig. 1, agent's behavior rules are as follows.

Rule 1. Choosing task executive way. If p_i undertakes tk_{uv}, knowledge need is:

$$K_{tk_{uv}} - K_i = [k_{tk_{uv}, i, 1}, k_{tk_{uv}, i, 2}, \ldots, k_{tk_{uv}, i, n}] \tag{6}$$

$k_{tk_{uv}, i, c} = k_{tk_{uv}, c} - k_{i,c}$, $c = 1, 2 \ldots n$. If $\forall k_{tk_{uv}, i, c} \leq 0$, p_i can finish tk_{uv} and do Rule 2. If $\exists k_{tk_{uv}, i, c} > 0$, p_i can't finish it and produce motivation of gaining c

kind knowledge. And $0 \leqslant k_{tk_{uv},i,c} \leqslant b_i$, p_i does Rule 3, otherwise, $k_{tk_{uv},i,c} > b_i$, p_i does Rule 2.

Rule 2. Executing task. Team increases the completed amount of task according to the amount of members has at each time. Team allocates a new task until performing task is finished. $F_{tk_u}(t)$ refers to the completed amount of tk_u at time t:

$$F_{tk_u}(t) = \sum_{i=1}^{M'} \sum_{k=1}^{N'} k_{i,k}(t) \tag{7}$$

M' is the number of members undertaking in tk_u and N' is knowledge numbers that tk_u needs at time t.

Rule 3. Knowledge interaction. Knowledge interaction process includes choosing interaction object, obtaining knowledge, memorizing interaction object's individual knowledge, generating new knowledge need, etc. Specific steps are as follows:

(1) p_i chooses interaction object who has the highest amount of knowledge and the largest knowledge similarity with himself. Results show that the higher knowledge similarity between members, the better knowledge interaction's efficiency and quantity. As $k_{tk_{uv},i,c}$, p_i chooses:

$$p_s = \max_j \left\{ u_{ijc} \times e^{s_{ij}-1} \right\}, \, j = 1, 2, \ldots, m \tag{8}$$

s_{ij} is knowledge similarity between p_i and p_j which represents as:

$$s_{ij} = \frac{1}{n} \sum_{k=1}^{n} \min \left\{ u_{iik}, u_{ijk} \right\} \tag{9}$$

(2) p_i obtains knowledge from p_s. $[k_{s,c}(t-1) - k_{i,c}(t-1)] > 0$ is the condition p_i interacts with p_s at time t. p_i obtains c kind knowledge as:

$$\begin{cases} k_{i,c}(t) = k_{i,c}(t-1) + \Delta k \\ \Delta k = k_{s,c}(t-1) \times a_i \times e^{min\{s_{is}, s_{si}\}-1} \end{cases} \tag{10}$$

at the same time, p_i and p_s memories as:

$$\begin{cases} u_{sic}(t) = k_{i,c}(t) \\ u_{isc}(t) = k_{s,c}(t) \end{cases} \tag{11}$$

$k_{tk_{uv},i,c}$ updates to:

$$k_{tk_{uv},i,c}(t) = \begin{cases} 0 & \Delta k \geqslant k_{tk_{uv},i,c}(t-1) \\ k_{tk_{uv},i,c}(t-1) - \Delta k & \Delta k < k_{tk_{uv},i,c}(t-1) \end{cases} \tag{12}$$

Rule 4. Knowledge forgetting. Knowledge and skills will be forgotten if they are not used for a long time. Forgetting curve is described as a curve close to exponential distribution, as shown in equation (13).

$$k_{i,c}(t) = k_{i,c}(t-1) \times e^{-c_i} \tag{13}$$

3.3 Team Knowledge Modeling

Definition 1. *Distributed knowledge is expressed as a three-dimensional knowledge space which is short for S_{DK}. As shown in Fig. 2(1), P(S) and P(O) axes stand for active and passive side in knowledge interaction, K axis stands for knowledge that team containing. $A(p_i, p_j, k_r) = dk_{ijr}$ means p_i knows "p_j has r kind knowledge as expertise" and the amount of r p_j has is dk_{ijr}.*

Definition 2. *Shared knowledge is described as a three-dimensional knowledge space, short for S_{SK}, shown in Fig. 2(2). In S_{SK}, $A'(p_i, p_j, k_r) = sk_{ijr} = sk_{jir} = min\{ck_{ijr}, ck_{jir}\}$ means p_i and p_j has r kind knowledge and the amount of r they shared is sk_{ijr}.*

Definition 3. *Team knowledge level is the average level of shared knowledge and distributed knowledge. It represents as:*

$$TKL = \frac{1}{2}(TDKL + TSKL) \tag{14}$$

TDKL is the level of distributed knowledge; TSKL is the level of shared knowledge. The formulas are as follows.

$$TDKL = \frac{1}{m(m-1)} \sum_{i,j=1}^{m} \left[\sum_{r=1}^{n} \frac{dk_{ijr}}{sgn(dk_{ijr})} \right] \tag{15}$$

$$TSKL = \frac{1}{m(m-1)} \sum_{i,j=1}^{m} \left[\sum_{r=1}^{n} \frac{sk_{ijr}}{sgn(sk_{ijr})} \right] \tag{16}$$

3.4 Running Steps of Computational Experiment Model

This model is utilized to integrate tasks, team knowledge, member's attributes and rules. It is driven by following procedure, simulating a knowledge intensive team which has M members, N kinds of knowledge performs U tasks orderly.

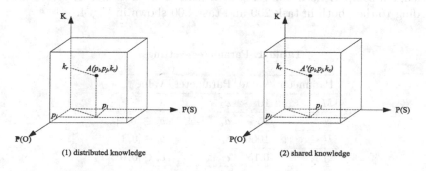

Fig. 2. Team knowledge

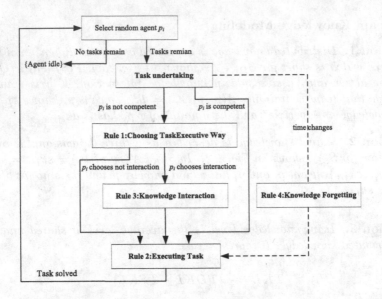

Fig. 3. Process of p_i completes a task

Step 1: Initialize the parameters of all members and tasks;
Step 2: Team performs a task and calculate *TKL*;
Step 3: Return to step 2, do the next task until all tasks are completed.
The process of each member completing a task is shown in Fig. 3.

4   Experiments and Results Analysis

The initial values of parameters in this model are shown in Table 1. In order to stabilize the experiment results, $U = 1000$.

In Fig. 4, it is showed that with the increasing tasks team knowledge is gradually formed, developed and tended to be a final stable state, representing a stable macro effect. Also we can observe that shared knowledge and distributed knowledge are experienced a development process from immaturity to maturity according to they both in task 200 and task 800 shown in Fig. 4.

Table 1. Parameters settings

Parameter	Value	Parameter	Value
M	10	ε_2	3
N	15	a_i	$a_i \in [0, 1]$
U	1000	b_i	$b_i \in [0, 1]$
ε_1	0.15	c_i	$c_i \in [0, 0.5]$

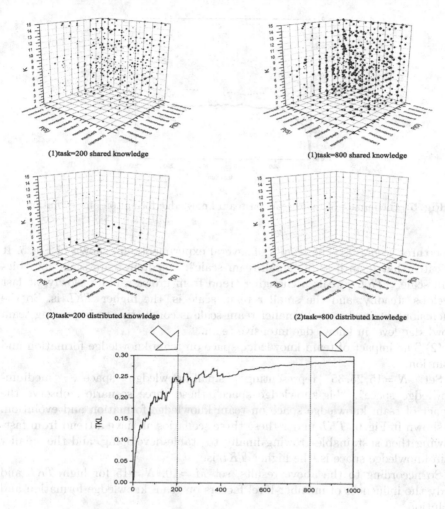

Fig. 4. Changes of team knowledge level

According to this computational model, factors affecting team knowledge formation and evolution can be divided into two categories, one is team-level factors like team scale M and team knowledge space N that is the knowledge numbers in knowledge intensive team; the other is member-level factors such as member's knowledge learning ability a_i, knowledge interaction willingness b_i and initial knowledge state K_i. So for each factor, set different values to describe different scenarios and compare team knowledge formation and evolution under scenarios, the factor's optimal value can provide reliable decision supports for managers to use team knowledge improving enterprise innovation.

(1) The impact of team scale on team knowledge formation and evolution

Set $M = 10, 15, 20$ standing for "small-team scale", "medium-team scale", "big-team scale" and observe team knowledge formation and evolution under

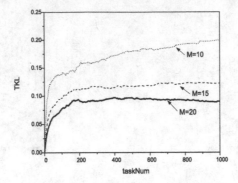

Fig. 5. The impact of team scale on team knowledge formation and evolution

these three scenarios. The results of several experiments are shown in Fig. 5. It indicates that regardless of "small-team scale", "medium-team scale" or "big-team scale", *TKL* reveals a changing trend from grows rapidly, slowly, at last develops steadily; and the smaller team scale is, the higher *TKL* is. So the conclusion obtained is that smaller team scale is conductive to improving team knowledge level in knowledge intensive team.

(2) The impact of team knowledge space on team knowledge formation and evolution

Set $N = 15, 25, 35$ representing "small-knowledge space", "medium-knowledge space", "big-knowledge space" these three scenarios, observe the impact of team knowledge space on team knowledge formation and evolution. As shown in Fig. 6, *TKL* under these three scenarios all have a trend from fast-growing then sustainable-growing finally to stable-developing, and the smaller team knowledge space is, the higher *TKL* is.

So according to the above results, set $M = 10$, $N = 15$ for high *TKL* and study the influences of member-level factors on team knowledge formation and evolution.

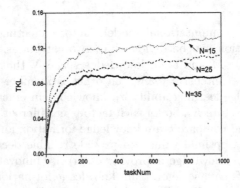

Fig. 6. The impact of knowledge space on team knowledge formation and evolution

(3) The impact of member's knowledge learning ability on team knowledge formation and evolution

Set $a_i = 0.1, 0.5, 0.9$ respectively representing "weak-knowledge learning ability", "moderate-knowledge learning ability", "strong-knowledge learning ability" to observe the influence of member's knowledge learning ability on team knowledge formation and evolution. The results in Fig. 7 show that TKL under "strong-knowledge learning ability" is highest and TKL under "weak-knowledge learning ability" is lowest. It indicates that the stronger member's knowledge learning ability is, the higher TKL is.

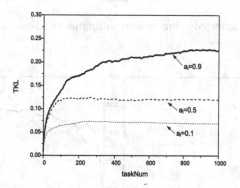

Fig. 7. The impact of knowledge learning ability on team knowledge formation and evolution

(4) The impact of member's knowledge interaction willingness on team knowledge formation and evolution

Set $b_i = 0.1, 0.5, 0.9$ respectively standing for "low-knowledge interaction willingness", "moderate-knowledge interaction willingness", "high-knowledge interaction willingness" to observe the influence of member's knowledge interaction willingness on team knowledge formation and evolution. As shown in Fig. 8, TKL under "low-knowledge interaction willingness" and "moderate-knowledge interaction willingness" has a, fast-growing, sustainable-growing then stable-growing, developing trend. However, TKL under "high-knowledge interaction willingness" grows rapidly firstly then declines subsequently at last develops stably. So in the long run TKL under "moderate-knowledge interaction willingness" is highest. This demonstrates that member's moderate-knowledge interaction willingness is conductive to team knowledge formation and evolution.

(5) The impact of member's initial knowledge state on team knowledge formation and evolution

Set member's initial knowledge state respectively representing "less", "medium", "more" and observe the influence of initial knowledge state on team knowledge formation and evolution. In Fig. 9, when member's initial knowledge state is under "less" scenario, TKL is lowest and unstable; when it is under "medium" and "more", TKL has a fast-growing developing trend, and it under

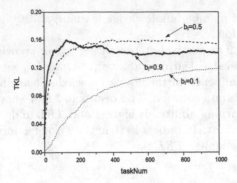

Fig. 8. The impact of knowledge interaction willingness on team knowledge formation and evolution

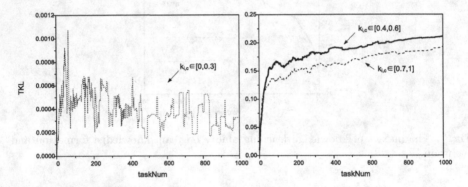

Fig. 9. The impact of initial knowledge state on team knowledge formation and evolution

"more" is higher than that under "medium". Thus the more member's initial knowledge state is, the higher TKL is.

5   Conclusion

In this paper, a computational experiment model of team knowledge formation and evolution is established, and the influences of team scale, team knowledge space, member's knowledge learning ability, knowledge interaction willingness and initial knowledge state on team knowledge formation and evolution are analyzed based on this model. Conclusions are obtained, including: the smaller team scale and team knowledge space are, the higher team knowledge level is; member's stronger knowledge learning ability, more initial knowledge state and moderate-knowledge interaction willingness are conductive to increase team knowledge level. Therefore, for a knowledge intensive team, smaller team scale and team knowledge space are beneficial to enhance innovation capabilities. And while member's knowledge learning ability is stronger, his initial knowledge state is higher and knowledge interaction intention is in medium level, it is advantageous to team innovation. The conclusions above can provide reliable decision supports for managers to use team knowledge improving enterprise innovation.

In the next step, we will use empirical research to carry out the subsequent research. On the one hand, model proposed in this paper can be modified and verified by comparing empirical results and simulation conclusions; on the other hand, this model can be better used in management practice. This is our main work in future.

Acknowledgments. This work is partly supported by the National Natural Science Foundation of China under Grant No. 71471028.

References

1. Chung, Y., Jackson, S.E.: The internal and external networks of knowledge intensive team. J. Manage. **39**, 442–468 (2013). doi:10.1177/0149206310394186
2. Lee, J.Y., Bachrach, D.G., Lewis, K.: Social network ties, transactive memory and performance in groups. Organ. Sci. **25**(3), 951–967 (2014). doi:10.1287/orsc.2013.0884
3. Sikorski, E.G., Johnson, T.E., Ruscher, P.E.: Team knowledge sharing intervention effects on team shared mental models and student performance in an undergraduate science course. J. Sci. Educ. Technol. **21**(6), 641–651 (2012). doi:10.1007/s10956-011-9353-9
4. Wildman, J.L., Thayer, A.L., Pavlas, D.: Team knowledge research: emerging trends and critical need. Hum. Factors. J. Hum. Factors Ergon. Soc. **54**(1), 84–111 (2012). doi:10.1177/0018720811425365
5. Kozlowski, S.W.Y., Ilgen, D.R.: Enhancing the effectiveness of work groups and teams. Psychol. Sci. Public Interest **7**(3), 77–124 (2006). doi:10.1111/j.1529-1006.2006.00030.x
6. Gang, Z.H., Jie, L.: Team mental models and transactive memory system: two ways of team knowledge representation. J. Dialect. Nat. **34**(1), 81–88 (2012)
7. Woolley, A.W., Malone, T.W.: Evidence for a collective intelligence factor in the performance of human groups. Science **330**, 686–688 (2010). doi:10.1126/science.1193147
8. Dechurch, L.A., Mesmer-Magnus, J.R.: The cognitive underpinnings of effective teamwork: a meta-analysis. J. Appl. Psychol. **95**(1), 32–53 (2010). doi:10.1037/a0017328
9. Johnson, T.E., Top, E., Yukselturk, E.: Team shared mental model as a contributing factor to team performance and students' course satisfaction in blended courses. Comput. Hum. Behav. **27**(6), 2330–2338 (2011). doi:10.1016/j.chb2011.07.012
10. Mancuso, V.F., Mcneese, M.D.: Effects of Integrated and Differentiated Team Knowledge Structures on Distributed Team Cognition. In: 56th Annual Meeting of the Human Factors and Ergonomics Society, pp. 388–392. SAGE Press, Boston (2012). doi:10.1177/1071181312561088
11. Nrico, R., Nchez-Manzanares, M.S., Gil, F., Gibson, C.: Team implicit coordination process: a team knowledge-based approach. Acad. Manage. Rev. **33**(1), 163–184 (2008). doi:10.5465/AMR.2008.27751276

12. Liao, J., O'Brien, A.T., Jimmieson, N.L.: Predicting transactive memory system in multidisciplinary teams: the interplay between team and professional identities. J. Bus. Res. **68**(5), 965–977 (2015). doi:10.1016/j.jbusres.2014.09.024

13. Chiang, Y.H., Shih, H.A., Hsu, C.C.: High commitment work system, transactive memory system, and new product performance. J. Bus. Res. **67**(4), 631–640 (2014). doi:10.1016/j.jbusres.2013.01.022

14. Lv, Y., Zhang, X., Kang, W.: Managing emergency traffic evacuation with a partially random destination allocation strategy: a computational experiment based optimization approach. IEEE. Trans. Intell. Transp. Syst. **16**(4), 2182–2192 (2015). doi:10.1109/TITS.2015.2399852

15. Monteiro, R.D.C., Ortiz, C., Svaiter, B.F.: An adaptive accelerated firstorder method for convex optimization. Comput. Optim. Appl. **64**(1), 1–43 (2016). doi:10.1007/s10589-015-9802-0

16. Poppenborg, J., Knust, S.: Modeling and optimizing the evacuation of hospitals based on the MRCPSP with resourse transfers. Eur. J. Comput. Optim. **4**, 1–32 (2016). doi:10.1007/s13675-015-0061-8

17. Long, Q.: An agent-based distributed computational experiment framework for virtual supply chain network development. Expert. Syst. Appl. **41**(9), 4094–4112 (2014). doi:10.1016/j.eswa.2014.01.001

18. Kozlowski, S.W.J., Klein, K.J.: A Multilevel Approach to Theory and Research in Orgaizations: Contextual, Temporal and Emerget Processes. Jossey-Bass, San Francisco (2000)

Towards a Service Value Co-creation Model for Older Adult Education

Jinfang Cai[✉] and Michitaka Kosaka

School of Knowledge Science,
Japan Advanced Institute of Science and Technology,
Nomi, Ishikawa, Japan
{caijinfang,kosa}@jaist.ac.jp

Abstract. With the raise of aging problem in many developed and even developing countries, Older Adult Education (OAE) has been considered an effective way to deal with the problem and widely adopted and deployed. It comes naturally another problem that how to fully utilize OAE, i.e., to maximize the value of it. Unlike business services to which numerous efforts are being made to achieve their maximal value, OAE is usually provided by government as a public non-profit service, and few attention has been paid to the maximization of its value. In this paper, we propose a concrete service value co-creation model that is specialized to OAE by studying two successful cases on OAE universities in Shanghai from service science perspective. The model is instantiated from a meta business service value co-creation model called KIKI model.

Keywords: Service · Value co-creation · Older adult education · KIKI model

1 Introduction

The nation is rapidly graying. Such statement is mentioned in a report [1] by American Council on Education, 2007. The aging problem also emerges even in developing countries such as China and could be worse than other countries due to its "One Family One Child" policy in history. The aging problem leads to various social problems such as workforce shortages, health and medical care burden and increasing demand for financial investment.

OAE is a process of knowledge creation, providing older adults with ways to improve the knowledge and skills and help them to live a better life in their third ages. It is first proposed by western countries as a special kind of post-secondary education for the older adult. Older adults mean those people aged 55 to 79 years, the prime years during which people are actively choosing how they will spend the third age of their lives [1]. Older adults are considered as resources to help address workforce shortages, solve community problems, mentor the next generation, and decrease the burden caused by aging population.

OAE is being increasingly adopted and deployed in many countries, and it is indeed helpful in some aspects such as to improve the life quality of retired older

© Springer Nature Singapore Pte Ltd. 2016
J. Chen et al. (Eds.): KSS 2016, CCIS 660, pp. 15–29, 2016.
DOI: 10.1007/978-981-10-2857-1_2

adults, to keep them both physically and mentally healthy, and to help them back to workplace. A number of studies show that older adults are positively engaged in educational programs [2]. However, based on our survey there is little literature on the maximization of the value of OAE, i.e., how to fully utilize OAE to make its value maximized. According to a survey from U.S. National Center on Education Statistics, only 27 % older adults prefer to attend work-related courses, and 21 % prefer personal interest courses [1]. Our survey in China also shows that older adults are more likely to attend those courses for their personal interest, but have little interest in work-related courses. OAE is well welcome to the older adult as a kind of social welfare, while governments provide such education to the older adult as a kind of investment. A problem arises that how to maximize the value of OAE for both the older learner and the education provider, i.e., to make both sides profitable from the education.

To address the aforementioned problem, the central question to answer is that what kind of education should be provided to the older learner so that the objectives of both the education provider and the older learner can be achieved. Although there does not exist a unique answer to the question due to the variety of the motivations of older learners and the challenges faced in deploying OAE, it is still desired to establish some model, by which one could develop step-wisely customized education systems for specific classes of older learners.

Unlike business services to which numerous efforts are being made to achieve their maximal values, OAE is usually provided by governments as a public service, and few attention has been paid to the maximization of its value. In this paper, we propose a concrete service value co-creation model that is specialized OAE from the perspective of service science. We consider the education for the older learner as a kind of service, and attempt to apply an abstract service value co-creation model called KIKI model to the education. KIKI model is originally proposed for the value co-creation of business services. In our work, we survey two successful Chinese OAE universities for older adults, analyze the development of their education system using KIKI model, and finally generalize a concrete model by instantiating each step in KIKI model. This study shows that KIKI model is also suited to value co-creation for OAE.

The rest of this paper is organized as follows. Section 2 presents a survey on the motivations and challenges in OAE. Section 3 views OAE from service science perspective and introduces KIKI model. Section 4 presents two case studies and the value co-creation model for OAE. Section 5 finally concludes the paper.

2   Motivations and Challenges in Older Adult Education

Motivation plays a central role in the process of older adult learning [8,11]. As described in the report [1], *at the heart of these challenges is the range of motivations and needs of the older adult population.* To maximize the value of OAE, it is necessary to balance the motivations between the older learner and OAE providers and hence to make the objectives of both sides achieved. In this section, we survey by literature review the motivations of the both sides and the challenges that OAE is facing in the reality.

2.1 Motivations of the Older Adult to Participating OAE

Increasing attention is being paid to the study of older adults' motivation to learn in OAE. In the work [11], Yin classifies the motivation of older adult learners into the following five kinds:

1. *Desire for knowledge*, i.e., learning new knowledge;
2. *Desire for stimulation*, i.e., having the chance to exercise their mind;
3. *Desire for self-fulfillment*, i.e., self-actualizing by pursuing a diploma;
4. *Desire for generativity*, i.e., taking responsibility for future generations by learning and teaching;
5. *Learning as a transition*, i.e., transiting to another life or career.

In the work [3], Dench and Regan summarize three major motivations that 74 % of older learners reported as very or fairly important to their study, including intellectual, personal and instrumental motivations. By intellectual motivation it means that learners want to increase their knowledge, to keep their mind active, to enjoy the challenge of learning new things. By personal motivation it means that learners want to gain qualifications for personal satisfaction and to take their life in different directions. Instrumental motivation means that learners want to learn for their work, to help their family, and to help with voluntary or community work. It is also reported that the third one is less important than other two motivations.

It can be apparently seen that motivations of older adults to higher education participation are diverse, which makes OAE differentiate itself from other educations such as youth school education.

2.2 Motivations of Offering Higher Education to Older Adults

The motivation (or aim) of providing higher education for the older adult is discussed in numerous literature from either theoretical or practical point of view. Theoretically, education is considered as a right that must be accessible throughout life, and therefore educational institutions become responsible for creating opportunities involving older adults themselves in activities backed up by the philosophy of lifelong learning and inter-generational education [4]. To be concrete, by higher education (a) it satisfies older adults' cultural needs; (b) it answers older adults' educational demands; (c) older adults contribute to their support through their work and tax payment.

Practically, the main objective of providing higher education to older adults is to alleviate or solve social problems caused by population aging. For instance, in the report [1], it is proposed that higher education for older adult can help address workforce shortages. Older adults' education experiences in Argentina emphasize the role that education has on empowering and learning new social roles or the re-signification of traditional ones [4]. In China, universities for retired cadres (those who take the leading or administrative role in government or state-owned sections) were founded since 1983 to answer their demands for keeping active and physically and mentally healthy. The University of the Third

Age (U3A), one of the most famous universities for older adults founded in France in 1972, formulates their three aims for the education of older adults [4]:

1. The first is around learning and "intellectual stimulation" – encouraging members to share their knowledge with others and learn from others;
2. The second is social and concerns providing social contacts for its membership, i.e., to provide continuing education, stimulation and companionship for retired people, as one committee member put out;
3. The third is an advocacy aim and asserts the ability of older adults to continue learning, i.e., to refute the idea of intellectual decline with age.

Benefits from OAE are multiple, not only to older adults but also to the society. Providers of OAE, e.g., colleges and universities for older adults may emphasize some of them, depending on the background of participants and the missions of founding colleges and universities.

2.3 Challenges in Higher Education for Older Adults

Challenges in higher education for older adults can be classified into two kinds. In one kind are the challenges for the older adult and in the other are the challenges for the higher education provider.

ACE identifies three barriers in the United States of America that could weaken the motivation of the older adult to learn. The three types of barriers include demographic barriers, attitudinal barriers and structural barriers [1].

1. *Demographic barriers*: Age, race and ethnicity, and geography are the main demographic variables that complicate decisions about higher education participation;
2. *Attitudinal barriers*: Both external attitude from advisers, family, and friends, and internal attitude from the older adult themselves can pose barriers for them to participate into higher education;
3. *Structure barriers*: Lack of transportation, support service and financing, and insufficient adaption of existing programs often keep older adults out of classroom.

Challenges for the higher education provider are mainly from shortage of funding, outreach barriers, and programming concerns [1]. Nowadays, OAE is still a public program that is mainly supported by government. By outreach barriers, it means that the educational needs and motivations of the older adult vary, as discussed in Sect. 2.1. Such variety forces colleges and universities to find means to reach older adults from different background, ranging from the PhD to high school dropout, from retried executives to part-time laborers, etc. Programming concerns mean the programs, courses and models that are adopted for older adults. Unlike school education, programs, courses, and models are various in both contents and forms.

Researchers do a lot of work on OAE which focus on theoretical research and practical research from education science perspective and social science perspective. Research contents consist of the impact and influence of OAE on society, development of senior education, education contents, education approaches, etc. Little attention has been paid to minimize the gaps of motivations between education providers and older learners. According to the situation survey and literature review, there are various barriers for realizing the service value of OAE for both the older learner and education providers. Some new solutions should be identified to maximize the value.

3 Older Adult Education in Service Science Perspective

As the rise in ascendance of the service sector, there has been an increasing interest by industry, government, and academia on understanding the determinants of productivity in service industries as well as innovation. In this section, we introduce the basic research of service science, and view OAE in service science perspective, and then propose a hypothesis of applying the value co-creation model, i.e., KIKI model, to OAE.

3.1 Service Science, Service System and Service Dominant Logic

Service science is the study of service system and of the co-creation of value within complex configurations of resources [10]. It has been becoming a common research thread in information and knowledge industries. More and more researches on service innovation and service value creation have been proposed and put forward. According to description of Sadahiko Oda, the chairman of a famous Japanese SPA hotel, service is an activity that provides professional techniques, satisfies the customer, and results in compensation for the service provider.

In service science, new concepts related to service have been proposed. Vargo and Lusch proposed the service dominant logic (SDL) concept which is different from traditional goods dominant logic (GDL), that the value is determined by customer on the basis of "value in use" [9,10]. In GDL, the determination of value is products based on concept "value in exchange". Whereas, in SDL, it leads to a new viewpoint of service through extension of its concept to include goods. The key point of SDL is that service value for customers is created through collaboration between service providers and customers, and the customer is a co-producer of the service and is an active participant of mutual interaction.

3.2 A Service Value Co-creation Model: KIKI Model

Value creation is the core purpose and central process of economic exchange [10]. A number of value co-creation models have been proposed in different logic such as Service Dominant Logic, Goods Dominant Logic, Experience based Service Value Co-creation, and Experience based Economy. A service value co-creation model can be either static or dynamic.

Kosaka et al. proposed a dynamic experienced-based service value co-creation process model called KIKI model [7], which has been successfully applied to many business cases such as Energy Saving Service System. The feature of KIKI model is that the concept of service field is applied to B-to-B (Business-to-Business) collaboration. Service field is a concept that is similar to electromagnetic field in physics, where electromagnetic power is determined by both the electric charge and the electromagnetic field where it is located. The value of a service is also determined by the service itself and the field which shows the context of provided service [6]. In KIKI model, the process model of service value co-creation consists of the following four steps:

Step 1: *Knowledge sharing in collaboration:* Service providers and receivers cooperate with each other, share their objectives and experience, and understand each other;

Step 2: *Identification of the service field:* Data collection and analyses are used to extract information of the first step, and identifying service field;

Step 3: *Knowledge creation for new service idea:* New service idea is designed based on the identified service field of step 2. Through participants' collaboration in value co-creation process, new knowledge of service is created by combining various service ideas and technologies;

Step 4: *Implementation of new service idea:* The created new idea is put into practice. Collaborators in service value co-creation process evaluate the results of knowledge creation step for the required service and take them into account in the following process for enhancing service.

These steps are iterated as a spiral development process. During the iteration of the process, the service field changes based on the change of the customer's experience in each step.

3.3   Service Value Co-creation Model for OAE

OAE is different from teenager education and professional education in many aspects, such as education forms and education objectives. The older learner have specific characteristics: (1) they expect to be treated with respect and recognition; (2) they want practical solutions to real-life problems; (3) they can reflect on and analyze individual experiences; (4) they are motivated by the possibility of fulfilling personal needs and aspirations; (5) they have different learning styles; and (6) they are capable of making their own decisions and taking charge of their own learning.

With the above facts, OAE can be viewed as a type of service, and it should be conducted according to SDL, in which older learners are active participants of the service. We propose research hypothesis as follows:

1. The value of OAE can be achieved through service value co-creation;
2. The old people are co-creators of the service value;
3. The model of service value co-creation structure can be used to analyze relationship between older learners and education providers;

4. The model of service value co-creation process (KIKI model) can be used to identify the service field of OAE, provide new services to older learners, and maximize their satisfaction and improve the value of education providers.

Based on above hypothesis, we propose the structure and process of service value co-creation model of OAE based on KIKI model, as depicted in Fig. 1.

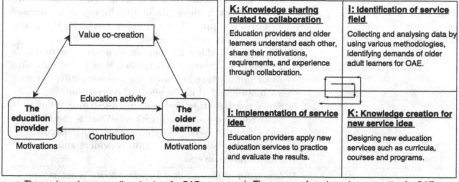

K: Knowledge sharing related to collaboration	**I: Identification of service field**
Education providers and older learners understand each other, share their motivations, requirements, and experience through collaboration.	Collecting and analysing data by using various methodologies, identifying demands of older adult learners for OAE.
I: Implementation of service idea	**K: Knowledge creation for new service idea**
Education providers apply new education services to practice and evaluate the results.	Designing new education services such as curricula, courses and programs.

a. The service value co-creation structure for OAE   b. The process of service value co-creation for OAE

Fig. 1. The abstract structure and process of value co-creation model for OAE from service science perspective

4 Case Study

In this section, we present two case studies to show how the service value co-creation model described in Sect. 3.3 can naturally depict the improvement process of the education service in two successful OAE universities in China, by which we say the hypothesis proposed in Sect. 3.3 is verified.

We choose two Chinese OAE universities mainly for two reasons: One reason is that OAE has been prevalently adopted in China along with the concept of Lifelong Learning (LLL) and Education for All (EFA) which have become central themes in education area. According to incomplete statistics, until 2013, the number of universities and schools for old people in China has dramatically increased to 43000 and learners have reached nearly five million [5]. The other reason is that the government plays one of the most significant roles in promoting elder learning. It gives the overall guidance and provides funding, venues and facilities. However, there is a gap between the rhetoric of national policy and the practice on the ground in China.

4.1 Design of the Survey Used in Case Studies

We use both qualitative and quantitative methods to collect data in two cases. The survey subjects include two groups, i.e., the education provider and the older

Table 1. Design of the survey subjects, methods and contents used in case studies

Subject	Method	Content
Education provider	Interview	1. *Education objective:* What is education objectives? What extent and how does it be achieved? 2. *Education situation:* what is overall education situation? What are provided education activities? What are students' response to the activities? 3. *Education content:* How to develop new education contents? What is principle of developing new courses? What is the process of designing new education contents? How to test new idea? 4. *The role of older learner in education activity:* Is there any co-creating experience? How to interact with older adult? In what ways do the university interact with students? What is effect of co-creation? 5. *Service innovation for senior education:* What is your feeling on senior education? What is the ideal older adult university in your mind? What is good relationship between education providers and education receivers?
The older learner	Questionnaire & Interview	1. *Learning objective and requirement:* What is your purpose of attending the university? What extent can your requirement be achieved? 2. *Learning experience and degree of satisfaction:* What have you learned from the university? Does the provided service meet your needs? 3. *Service co-creation experience:* Do you have experience on giving your suggestions to education providers? Do you think it is necessary for the learners to join the reform? Will you positively get yourself into the reform? 4. *Service innovation for senior education:* Do you think the current senior college should be reformed? What should be improved? Do you have any ideas? What is the ideal older adult university in your mind? What is good relationship between education providers and receivers?

learner, who are two major groups in education system. We conduct survey in order to gain deep understanding of the situation of each university such as the objectives of founding the universities for the education providers, the objectives of participating the education for the older learners, and the satisfaction degree of the both two groups to the current education. We collect data from them and analyze it using the SPSS tool.

Table 1 shows the sketch of the design of our survey approach used in the two case studies. We survey the education provider mainly by the means of face-to-face interview, and the older learner mainly by questionnaire and interview. The questionnaire consists of 24 questions which are divided into three different types, i.e., motivations and needs, satisfaction, and suggestions. We also interview some older learners such as class monitors who usually are more motivated and positive to participate education activities.

4.2   Case I: Shanghai University for the Retired Veteran Cadre

Shanghai University for the Retired Veteran Cadre was founded in 1985 by Shanghai local government, and developed rapidly to be a national advanced older adult university and a demonstrative university in Shanghai. It has more than 3000 students who are once leaders retired from government departments. It provides more than 40 types of inner-classroom courses to students. In addition, it has outstretched education activities, such as learning salons and student associations. The government gives financial support, objectives, and administrative guide to the university. The administrators and working staff deploy education activities, and provide them to retired veteran cadres.

Data Collection and Analysis. On the older learner side, we collect 100 pieces of questionnaire among students randomly and interview three student representatives. We investigate their purposes of attending the university and satisfaction degree with provided education service from 7 aspects(the overall degree of satisfaction with served education, reasonality of courses, appropriateness of teaching methods, dedication of teachers, scientificity of teaching contents, satisfaction with infrastructure, and helpfulness of learning activities).

Figure 4 shows the distribution of the purposes of attending to the university and satisfaction degree with provided service. There may be multiple purposes for older learners to attend to the university. The first three purposes include *to learn and enrich lives, to keep physically and mentally healthy* and *to meet hobbies*, while the percentage of the learners who have other purposes such as *to make friends, to keep pace with society* and *to serve society* are quite low, i.e., under 35 %. Figure 4(b) depicts the satisfaction of the learners with the education provided by the university. More that 90 % older respondents are satisfied with the current education in seven aspects, although they think there are still space to improve the education. These data reflect the university conducts successful education activities from the older learners' point of view.

On education providers' side, we try to explore the mechanism which makes its education successful in practice. We interview an education administrator, a researcher, and a lecturer based on the investigation design shown in Table 1. We conclude the successful experience by using qualitative method.

The interviewed administrator said that "*the main objective of the university is to make it a base to advocate innovative theory of the party, to spread new knowledge, to study Chinese traditional culture, and to make examples of retired veteran cadres*". The researcher and administrator emphasized that "*older learners are the first consideration factor of all education activities*" in their education system. On the development of new course and service idea, the researcher said that it is mainly based on the needs and interests of older learners. They also design some new courses based on the objectives of university and then give some directions to learners to help them develop their interests. Besides, some

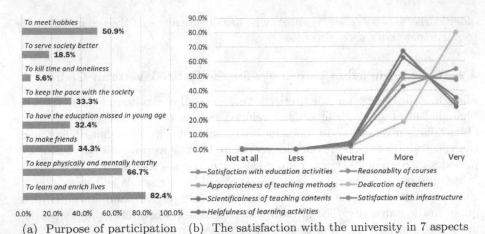

(a) Purpose of participation   (b) The satisfaction with the university in 7 aspects

Fig. 2. The survey result on purpose distribution of older learners and their satisfaction degree with education service in Case I

state-of-the-art courses like computer skills are provided to help older learners to keep pace with the society.

The university develops multiple ways to interact with the learners. Older learners are organized by classes, and in each class a monitor is assigned, who is mainly responsible for recording the feedback of his classmates every day. Such record is called *class diary*. These class diaries are the main data source for university administrators to analyze their education quality. At the end of each term, the university also organizes meetings for class representatives, monitors and administrators to listen to their opinions with each other.

The surveyed researcher said: *the process of developing our education system consists of following several steps: Investigation on the requirement of students through formal and informal forms, like questionnaire, face-to-face conversation, etc. Our research teams collect the data, conclude the effective information and design some new education ideas, and then conduct the second investigation to get the overall attitude of students towards new ideas. If the percentage of acceptance is high, then the idea can be tested by using seminar four times a term. We will collect students' interests again at the end of each term. If the result is good, the course can be applied as a regular course. The university evaluates the course by using our own evaluation system at the end of the first year of it.* An evaluation system is developed in the university to check if a newly opened course indeed achieves its objectives. The above statements show that the education providers interact with the students frequently in many ways and take their needs as the main factors to determine the courses.

A Refined Service Value Co-creation Model for OAE in Case I. In the case, the university conducts a type of need-oriented education. The older students' needs play central roles in service consideration. From our aforementioned

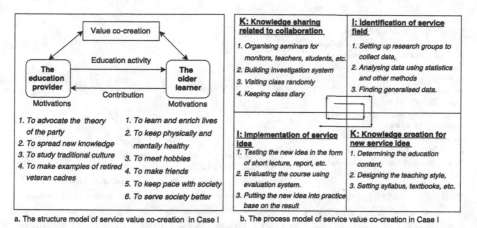

a. The structure model of service value co-creation in Case I   b. The process model of service value co-creation in Case I

Fig. 3. The structure and process model of value co-creation for OAE in Case I

survey, we can divide the process of establishing a successful OAE system into four steps: (1) collaborating and sharing knowledge with the older learner through formal and informal ways; (2) professional research teams collect and analyze data that are important to course designing; (3) new education service is designed based on analysis result and (4) new service is evaluated.

Through comparing research hypothesis with the practice of case I, we found that the hypothesis in Sect. 3.3 are suitable for OAE. The service value of OAE is a co-created process, older learners are co-creators of the value. Apparently, such process conforms to KIKI model.

We establish a concrete service value co-creation model for OAE in Case I, derived from KIKI model with the objective domains of both the education provider and the older learner being identified, and each step in the process is refined. Figure 3 shows the refined model. In the structure model, we identify four main objectives of the education provider and six main objectives of the older learner, as shown in Fig. 3(a). In the process model, concrete approaches to achieving each step is summarized based on our survey, as shown in Fig. 3(b).

4.3 Case II: Shanghai University for the Elderly

Shanghai University for the Elderly was also founded in 1985, aiming at *being spiritual home of old people and getting high satisfaction of student with the running of university*. It provides 109 kinds of course to nearly 10000 older students, and has no enrollment limitation compared with Case I.

Data Collection and Analysis. We survey both the older learner and staff in the university and collect data as we do in Case I.

On the older learner's side, we collect 112 pieces of questionnaire among students randomly and interview three student representatives. We invest their

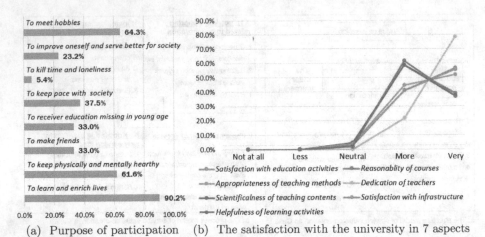

(a) Purpose of participation   (b) The satisfaction with the university in 7 aspects

Fig. 4. The survey result on purpose distribution of older learners and their satisfaction degree with education service in Case II

purposes of attending the university and satisfaction degree with provided education service from 7 aspects. Figure 4 shows the main purposes and needs for the elderly attending OAE university. The first purpose is *to learn and enrich lives* which is 90.2 %. The percentage of choosing *to meet hobbies* and *to keep physically and mentally healthy* is also very high, they are 64.3 % and 61.6 %. While the percentage of other purposes such as *to keep pace with society*, *to make friends*, *to receive education missing in young age* and *to improve oneself and serve better for society* are quite low. Over 90 % respondents are satisfied with the provided education service.

On the side of education provider, we interview four education providers for their experience of developing education service, they are education administrator, dean, a part-time lecture and a Third Class leader. We also participate some Third Classroom activities to obtain more detailed data.

The interviewed administrator said that the development of new course and service is mainly based on the needs and interests of older learners. They also explore some unknown needs of learners and lead them to learn. These needs may be unconscious to old people but they are really interested. Some state-of-the-art courses like photo shop course are provided to help older learners to keep pace with the society. The leader of Third Classroom gave details about the channel of teaching *"we have three classrooms: First Class, Second Class and Third Class. The First Class is a traditional one; the Second Class is called club or after-school activity which is organized by students who have the same interests. The Third Class can be named social practice, which is an important way for student to make contributions to community and society."*

The university lays emphasis on the comments and reflect of student on teaching activities and learning experience. The dean said *"the main method we used is monitor responsibility system under the leadership of dean. Every classroom*

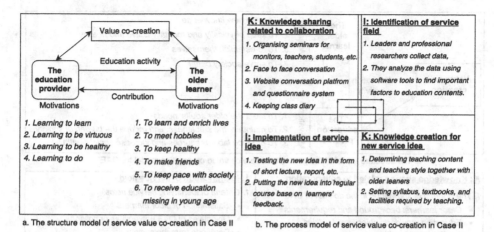

Fig. 5. The structure and process model of value co-creation for OAE in Case II

keeps a dairy note to record the detail of student information, needs, questions and the Monitor transfers those details to the dean. There are also seminars for monitors to share their opinions. Moreover, the forum of teacher representatives and student representatives are held twice a term to get more detailed information of provided service. BBS is also accessible for learners to reflect their problems anytime." The lecture told that *"the course system we offered looks like a supermarket and buffet service that suits the different requirements and characteristics.".* An evaluation system is developed in the university to check if a newly opened course indeed achieves its objectives and also be used to evaluate teaching quality of lecturers.

A Refined Service Value Co-creation Model for OAE in Case II. The above statements reflect that the education providers interact with the older students frequently in many ways and take their needs as the main factors to develop new services, deploy them, evaluate them, and improve them, which apparently conform to the four steps in KIKI model.

Figure 5 shows the concrete model for Case II. In the structure model, objectives of the service provider and the older learner in the university are identified. In the process model, we show the means of how the provider and the learner interact in each step to develop new courses and activities.

4.4 A Generalized Value Co-creation Model for OAE

From the survey and analysis result described in Sects. 4.2 and 4.3, we can find that the processes of developing their education service system in two universities conform to KIKI model, although the two universities are two different types of OAE universities. For instance, the university is Case I is opened only for a specific class of older learners, while the one in Case II is opened for any older

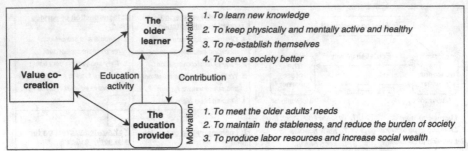

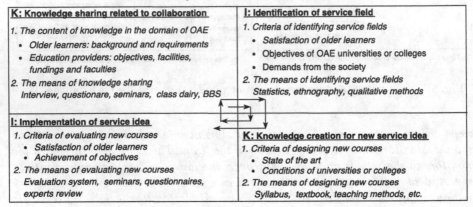

a. The structure model of the generalised service value co-creation model for OAE

b. The process model of the generalised service value co-creation model for OAE

Fig. 6. The structure and process of the generalized value co-creation model for OAE

learners. The needs of old learners in Case II are more varied than those in Case I. The older learners in Case I usually have rich education experience at universities and work experience at government, and hence their needs are sometimes more difficult to meet than those in Case II.

By unifying the two concrete value co-creation models in Case I and II, we generalize a value co-creation model for more general cases of OAE, as depicted in Fig. 6. In the model, we identify the contents and the means of knowledge sharing, and define the criteria and means of identifying service field, design new service, and implement the new service in the domain of OAE. The model is an instantiated one from KIKI model, but more abstract than those in the two cases in that the criteria and means in the model are more generic to OAE.

5   Conclusion and Future Work

In this paper, we presented a generalized service value co-creation model for OAE. The model is derived from a meta business model called KIKI model. By studying the process of developing education systems in two Chinese OAE universities where the older learners have high satisfaction, we identified the content and approaches to knowledge sharing, and generalized the criteria and approaches in each step of the process model. By applying the model, we can establish customized education services and activities that can meet both the older learner and education provider's objectives. In this way, the values of these customized services and activities are maximized.

Regarding to the future work, we are considering conducting more case studies on OAE universities or colleges not only in China but also in other countries, by which we can make the content of knowledge sharing in the domain of OAE, the criteria and approaches in each step in the model more generic, resulting in a more general model for the value co-creation of OAE. By conducting more case studies, we could also examine the generality of the proposed model by checking whether it is suited to other cases in different countries.

References

1. Reinvesting in the Old Age: Older Adults and Higher Education. American Council on Education (2007)
2. Danner: Late-life learners at the university: the donovan scholars program at age twenty-five. Educ. Gerontol. **19**(3), 217–239 (1993)
3. Dench, S., Jo, R.: Learning in later life: Motivation and impact. Research report RR183 (2000)
4. Findsen, B., Formosa, M.: International Perspectives on Older Adult Education Research, Policies and Practice, vol. 22. Springer, Switzerland (2016)
5. Gu, X.: Zai guoji laonian daxue xiehui lishihui ji quanti huiyi shang de zhici [opening address on the plenary meeting of international association of universities of third age]. Laonian jiaoyu (Laonian daxue), pp. 8–9 (2013)
6. Kosaka, M.: A service value creation model and the role of ethnography. INTECH Open Access Publisher (2012)
7. Kosaka, M., Zhang, Q., Dong, W., Wang, J.: Service value co-creation model considering experience based on service field concept. In: 9th International Conference on Service Systems and Service Management, pp. 724–729. IEEE (2012)
8. Mulenga, D., Liang, J.S.: Motivations for older adults' participation in distance education: A study at the National Open University of Taiwan. Int. J. Lifelong Educ. **27**(3), 289–314 (2008)
9. Vargo, S.L., Lusch, R.F.: Evolving to a new dominant logic for marketing. J. Market. **68**(1), 1–17 (2004)
10. Vargo, S.L., Maglio, P.P., Akaka, M.A.: On value and value co-creation: a service systems and service logic perspective. Eur. Manag. J. **26**, 145–152 (2008)
11. Yin, Y.Y.: Older adults' motivation to learn in higher education. In: The 19th Annual African Diaspora Adult Education Research Pre-Conference, pp. 764–770 (2011)

Knowledge Sharing via Informal Communities in a Globally Distributed Organization

Penny Chen, Yen Cheung, Vincent C.S. Lee[⊠], and Adam Hart

Faculty of IT, Clayton Campus, Monash University,
25 Exhibition Walk, Clayton, Australia
pjychen@gmail.com, adhart81@gmail.com,
{yen.cheung,vincent.cs.lee}@monash.edu

Abstract. This research studies the geographic diversity and variation in informal communities within a globally distributed bank in Australia and how these differences impact on the effectiveness of internal collaboration within a Bank. Its social networking technologies with implications for their deployment and use are also explored. Online survey data is collected from 373 employees across all its main branches in Australia and Asia. Three main findings are: collectivist-oriented employees tend to feel more positive in terms of their work outcomes after informal community interactions than individualist-oriented employees; individualist-oriented employees are slightly less willing to share knowledge in informal communities than collectivist-oriented employees and e-mail is the most commonly used and preferred ICT tool for communication in informal communities across all global offices, regardless of cultural orientation. Suggestions for further work include doing a study on organisational culture versus geographic culture or doing a study on organisations in transitioning geographic cultures.

Keywords: Knowledge sharing · Informal communities · Culture · Individualism · Collectivism · Socialization

1 Introduction

Enterprises today have strategic and operational goals to gain global market presence. Hence knowledge sharing in the course of performing work related activities is no longer characterised by rigid hierarchical organisational structures, functional specialisation, and associated communication processes and ICT tools. Globalization wave drives the need for global distribution of offices to cope with increasing customer demand for higher quality of service and shorter product development life cycle. Advancements in ICT have dramatically made it possible to take on the challenges and changed the way in which we work. Recent body of research literature [4] postulates that efficiency and effectiveness of the work force can be improved through relax formalities in communicating and sharing of knowledge in the workplace and beyond the work context. This presents a challenge for senior management in the home office to implement effective policies for its employees across all its global offices. Due to differences in cultural backgrounds, management needs to be culturally agile to

© Springer Nature Singapore Pte Ltd. 2016
J. Chen et al. (Eds.): KSS 2016, CCIS 660, pp. 30–43, 2016.
DOI: 10.1007/978-981-10-2857-1_3

employees especially those offshore employees to ensure the success of its global offshores. One of the most important cultural or geographic "dimensions" that explains differing behaviours is the individualism-collectivism dichotomy. At the same time, cultural dimension can also affect how employees interact with ICTs for communication purposes and this will also be explored in this research.

This research studies the geographic diversity and variation in informal communities within a globally distributed bank in Australia and how these differences impact on the effectiveness of internal collaboration within a Bank. Its social (informal) networking technologies with implications for their deployment and use are also explored.

1.1 Background

The organization chosen in this study has wholly-owned subsidiaries operating mainly in Asia, making it a captive operation. The wholly owned subsidiaries with the largest number of employees are located in China, Fiji, India and The Philippines and these will be the focus of the study. One way to understand the cultural differences between a head office and its overseas captives in an organization is to research how the different communities in each global office communicate.

According to [19], culture is a system of beliefs embedded within a society and affects people's behaviours and the way they communicate within an organization. [10] believes that employee behaviour is affected by the societal value system they are from. [11] creates a cultural index to explain how certain cultural dimensions affect the knowledge transfer process. According to [25], the most important dimension of cultural differences is the relative emphasis on individualism and collectivism (IC). Individualism/collectivism is "the extent to which a person sees himself or herself as an individual rather than as part of a group" [5]. How does the IC cultural dichotomy affect the way in which employees share knowledge? According to [11], individualist-oriented employees tend to think before they share knowledge, "How does this benefit me?" whilst collectivist-oriented employees tend to think, "How will this benefit the group?" Based on these ingrained cultural make ups, the following differences in how different cultures of employees manifest themselves in the knowledge transfer process are shown in Table 1 below.

The ways in which employees decide to share knowledge is affected by whether or not they are working in an individualist-oriented cultural region or a collectivist-oriented cultural region as seen in Table 1 below. Due to these differences in communication

Table 1. Individualist-oriented employees vs. Collectivist-oriented employees according to Hofstede's (2001) cultural index

Individualists	Collectivists
'How does this benefit me?'	'How does this benefit the group?'
Less likely to divulge knowledge	More likely to divulge knowledge
Need individual incentives	Need group incentives
Less relationship-dependent	More relationship-dependent
Better at sharing explicit knowledge	Better at sharing tacit knowledge

styles, the knowledge transfer process is different, and this will affect what type of ICT tools individualist-employees prefer to communicate with in comparison to the type of ICT tools collectivist-employees prefer to communicate with [12, 16]. As seen in [11]'s cultural index in Table 1, individualist-oriented employees are better at sharing explicit knowledge, which according to [16], yields higher preference for telephone, e-mail and repositories as ICT tools for communication. On the other hand, collectivist-oriented employees are better at sharing tacit knowledge, which according to [16] yields higher preference for telephone, e-mail and instant messenger as ICT tools for communication.

Asian countries, particularly China, tend to display more collectivist qualities [6]. On the other hand, western culture is seen to be more individualistic in nature [6]. In this study, the organisation has offices in both individualist-oriented cultures (Australia) and collectivist-oriented cultures (China, India, The Philippines and Fiji). This set-up of the organisation gives rise to researching how the IC dichotomy affects how employees collaborate and share knowledge in a captive operation. Thus the main research question for this study is:

How does geographic diversity affect internal collaboration within a large globally distributed organization with a focus on the deployment of ICT networking technologies?

Based on the background information on the role of culture on knowledge transfer processes and communities, the following sub-questions were addressed in this research:

1. Do the cultural dimensions of "individualism" or "collectivism" have an effect on the outcomes of informal communities in relation to employee work performance (i.e. the efficiency and effectiveness)?
2. How do the cultural dimensions of "individualism" and "collectivism" affect the willingness of employees to share their knowledge with other employees in work informal communities?
3. Does the dichotomy of "individualism-collectivism" affect the frequency of employee usage of ICTs and their preferences for certain types ICTs used for informal communication?

2 Literature Review and Hypotheses

2.1 The Dichotomy of Knowledge Transfer Processes in Individualist versus Collectivist Geographies

An interesting body of research that has emerged from captive organisations is how similar or different the knowledge transfer process is between the home office and its subsidiaries overseas due to different cultural dimensions. In an organisation, knowledge is seen as being embedded in the organisation's rules, routines, structures, technologies and employees [7]. The knowledge transfer process involves a provider and a receiver and describes the method of the partial or identical replication of knowledge from one place to another [15]. When discussing knowledge transfer, it is important to consider these three dimensions: type, embodiment and transformation [18]. [18]

believes one needs to know if the knowledge is initially embedded in a person or in a process, if the knowledge is tacit or explicit and how it is being transferred. Tacit knowledge is knowledge embedded in procedures, routines, culture, values and ideals whilst knowledge that can be captured in sentences and recorded in artefacts is explicit [13]. Since tacit knowledge is embedded, it is more complex and harder to decipher and hence the transfer of this type of knowledge requires personal and deeper connections between the provider and receiver [3].

The geographic environment of an office may have a significant impact on this knowledge transfer process. Communities, by definition, are bound by informal relationships that share a common context (Lesser and Prusak, 1999). In the context of this research, informal communities are social interactions that happen outside the formal work reporting structure. Inkpen and Tsang (2005) claim that social interactions in informal settings in the workplace are considered rich sources of knowledge. The pattern of communication in informal communities may differ across different global regions. This could be due to the cultural differences between global regions [18]. According to [19], culture is a system of beliefs embedded within a society and affects their behaviours and the way they communicate within an organization. [10] believes that employee behaviour is affected by the societal value system they are from. Hence, it would be noteworthy to explore how "cultural" or "geographic" dimensions affect the way in which knowledge is transferred in informal communities within a multinational corporation whose subsidiaries are located in heterogeneous cultures like the financial organisation in this research.

2.2   The Cultural "Dimension" of Individualism and Collectivism

Individualism/collectivism (IC) is the "extent to which a person sees himself or herself as an individual rather than part of a group" [5]. This geographic dimension concerns itself with studying how self-interested individuals in one geographic culture are in comparison to another culture. The IC dimension is important when studying knowledge transfer because it determines if a community with more self-interested individuals affect how much knowledge is transferred and to whom the knowledge is transferred [18].

People in more individualistic cultures have the tendency to be more self-interested and care about their interests over that of a group [9]. Hence, an individualistic culture is usually characterized by loose relationship ties because of the existence of more self-interested individuals [5]. In this culture, employees are looking out for themselves and before they approach a task they always ask themselves the question "how does this benefit me?". Individualist employees are more calculative [12] with their involvement in communities and have a strong need for freedom. In terms of knowledge exchange, individualist employees are less likely to divulge information with others in their communities unless this action will benefit them in some way [5]. This is because they believe knowledge is the property of individuals rather than the company [14]. Furthermore, [3] claims that individualist employees have a higher preference for explicit knowledge over tacit knowledge as they are less relationship-dependent. Based on these observations and with offices located in geographies characterized by a more individualistic culture, the Financial organization is more likely to have informal communities with less knowledge transfer but more exchange of explicit knowledge.

Collectivist cultures are driven more by group interest rather than self-interest [5]. According to [5], collectivist individuals believe that the concept of self-interest should play a secondary role to a group's overall interest. This could work in favour of a firm because employees care more about the company's goals than their own professional goals and hence will divulge as much information as they know if it benefits the overall goals of the firm [14]. With regards to communication in informal communities that are in subsidiaries operating in collectivist geographies, this "group mentality" dimension could manifest itself as having employees being more honest and open in their conversations rather than calculative. Collectivist employees are also better at transferring tacit knowledge because they tend to have closer relationships with their colleagues and have a higher preference for face-to-face conversation over individualist employees [12]. Hence based on these observations a Financial organization, characterized by a more collectivist culture, is more likely to have informal communities with more knowledge transfer as well as more exchange of tacit knowledge.

The above discussion leads to the first hypothesis for this study:

H_{o1} :   **There is no difference between the outcomes (effectiveness/efficiency) of knowledge sharing between informal communities in individualist-oriented Geographies and informal communities in collectivist-oriented Geographies.**

H_{A1} :   **The Bank's global offices embedded in more collectivist-oriented Geographies will benefit more from their informal communities than those in more individualist-oriented Geographies because there is an alignment between collectivist values and knowledge sharing.**

2.3   Countries That Display Individualism and Collectivism

Whilst not one country is purely collectivist or individualist, some countries display more attributes of one or the other along the cultural spectrum. Asian countries, especially China, tend to display more collectivist qualities [6]. Individuals in Asian countries tend to be more group-oriented and less aggressive when working with others in order to strengthen their interpersonal relationships [25]. [20] finds that individuals from Chinese culture are more accustomed to knowledge sharing because they have a practice of passing down wisdom from family to family through the generations. On the other hand, Western culture is seen to be more individualistic in nature [6]. In Western countries like the USA and Australia, individuals are seen to display more individualistic cultural characteristics like being more self-interested in their own outcomes rather than a group outcome and hence, are more aggressive in group situations at the expense sometimes of interpersonal relationships [17]. However, the line between collectivist Asian countries and individualist Western countries is not clearly drawn because some Asian countries like Malaysia imitate Western culture in their everyday lives [20]. This above research leads to the second hypothesis given below:

H_{o2} : There is no difference between informal communities in individualist-oriented Geographies and informal communities in collectivist-oriented Geographies with regards to willingness of employees to share their knowledge with other employees.

H_{A2} : Employees in more collectivist-oriented Geographies, such as Asia, are more willing to share their knowledge with their colleagues than employees in more individualist-oriented Geographies like Australia.

2.4   The SECI Model, ICTs and the Individualist-Collectivist Dichotomy

The socialization, externalization, combination and internalization (SECI) model is often used to describe how knowledge conversion happens in a group setting [21]. For knowledge transfer in informal communities, the most relevant aspects of this model are socialization and combination. Socialization is the process of developing new knowledge through shared experiences and involves the transfer of tacit to tacit knowledge [23]. Combination involves the conversion of explicit to explicit knowledge and provides a form of knowledge repository that serves as a resource for others [22]. Individualist-oriented employees tend to prefer explicit knowledge communication than tacit knowledge communication because they carry themselves with more autonomy in the workplace and are less relationship-dependent [3, 12]. On the other hand, collectivist-oriented employees were better at transferring tacit knowledge because they tend to have closer relationships with their colleagues and have a higher preference for face-to-face conversation over individualist employees 16] surveyed a sample of employees engaged in both types of processes in the SECI model and found which mix of ICT tools were preferred for each process. Collectivist-oriented companies like Japanese companies may emphasize the importance of the Socialization mode, because they see "sharing and creating tacit knowledge through direct experience" p. 9 of [28] as essential for successful knowledge creation. Companies from the West which are more individualist-oriented are more likely to focus primarily on the Combination mode, as this is strongly about explicit knowledge and about "systemizing and applying explicit knowledge and information" p. 9 of [28].

As mentioned in a survey conducted by [26, 27] suggests that collectivist-oriented groups do favour socialization as their preferred mode of communication and western countries that are more individualist-oriented have a stronger focus on combination as a preferred mode of communication in the [21] SECI model. As collectivist-oriented employees prefer to share tacit knowledge [11] which is described by the process of socialization, it was found that telephone, e-mail and instant messenger were the most frequently used tools for this process in a survey done by [16]. As individualist-oriented employees prefer to share explicit knowledge [11] which is described by the process of combination, it was found that telephone, repositories and e-mails were the most frequently used tools for this process [16]. From the above literature, it could be said that individualist-oriented employees may have a preference for repositories and e-mail, where explicit information is shared. And collectivist-oriented employees may have a stronger preference for telephone and e-mail, where tacit knowledge is shared [16].

The above information leads to the third hypothesis for this study:

H_{o3} :   **There is no difference between ICT preferences and frequency of use by employees in informal communities in individualist-oriented Geographies and informal communities in collectivist-oriented Geographies.**

H_{A3} :   **There are different preferences and uses (email, social network, instant messenger, phone, work depositories) for the ICT tools in informal communities across the different global hubs due to the aforementioned collective-individualistic dichotomy.**

3   Research Design

3.1   Data Collection

Online surveying is chosen as the way to collect data from employees of different global hubs of the bank. Some of the main advantages of using online questionnaires for research include having access to unique populations which would not be possible to reach through other channels i.e. overseas offices [29]; saving time with distribution of the survey and collection of answers [30]; and saving costs by moving to an electronic medium from a paper format [1]. Qualtrics is used as the online survey software of choice to administer the questionnaires because it was the main site the Bank used to perform its own surveys. The online survey questions were designed using a 7-point Likert scale (Strongly disagree, disagree, somewhat disagree, neither, somewhat agree, agree, strongly agree). The Likert scale is often used in surveys to help yield numerical data from respondents for measurable purposes later on in the research [2].

Due to time restraints, the survey in this research was sent out in two rounds and was open for one week to each global hub. In Table 2, the number of surveys sent out to each global hub is shown, as well as the number of actual responses collected and the response rate of each global hub.

Table 2. Survey response rate for each global hub of the Bank

Country	Total number of surveys sent out	Actual number of responses received	Response rate
Australia	491	37	7.54 %
India	1952	230	11.78 %
China	118	70	59.32 %
Fiji	75	23	30.67 %
The Philippines	352	13	3.70 %

The fact that the sample sizes for each global hub were unequal and the fact that some of the countries do not meet the minimum sample size requirement for their respective populations is a slight limitation to this study. However, the data analysis technique that is used in this study (as described in the next section) helps to adjust sample sizes to make more accurate and *statistically significant* comparisons between the global hubs.

3.2   Data Analysis

Statistical analysis using the one-way analysis of variance (ANOVA) test is used to compare each country to prove/disprove the 3 hypotheses in this study. The test is performed on the IBM SPSS Statistics (SPSS) as this is the most widely used computer package used in social science for quantitative analysis [24].

The one-way ANOVA is used to determine whether there are any significant differences between the means of three or more independent (unrelated) groups [24]. The one-way ANOVA compares the means between the groups concerned and determines whether any of those means for the groups are significantly different from each other. The one-way ANOVA was chosen over the t-test to compare the global hubs in this study because the t-test is more suited to comparing the results of two groups, whereas the one-way ANOVA is used to determine whether there are differences among more than two groups, and it reduces the chance of error in calculations from performing multiple t-tests [8].

Specifically, the one-way ANOVA tests the null hypothesis:

$$H_o : \mu_1 = \mu_2 = \cdots = \mu_k$$

Where μ = group mean and k = number of groups. If the one-way ANOVA returns a significant result, we accept the alternative hypothesis (H_A), which shows there are at least 2 group means that are significantly different from each other. In this study, μ = mean response from the survey and k = number of countries surveyed in this study $k = 5$. The mean for each country's response in the survey could be calculated numerically due to the fact each statement relating to a hypothesis is linked to a numerical value response, as structured from the Likert scale design of the survey.

3.3   Post-hoc Tests

The one-way ANOVA test shows whether there is an overall difference between groups, but it does not tell which specific groups differed - post hoc tests do [8]. As post-hoc tests are used to confirm where the differences occurred between groups, they should only be run when there has been an overall significant difference in the groups' means (i.e., a significant one-way ANOVA result). Post-hoc tests attempt to control the experiment error rate, [24] claims in social research this error rate is usually 0.05. Post-hoc tests are termed a posteriori tests, i.e. performed after the event. In this study, post-hoc tests were performed when the one-way ANOVA test result was statistically significant i.e. above the alpha level of 0.05 indicating that there was a significant difference between two or more of the countries' mean to a statement in the survey. In this study, the post-hoc tests were used to help determine which countries differed from the others in their response to a statement e.g. was Australia the stand out anomaly to a certain statement about willingness to share knowledge? In this study, the post-hoc test used was the Tukey procedure, which is a multiple comparison test that tests a comparison of all possible pairs of means [8]. The Tukey procedure also helped consolidate the unequal sample sizes for each sample group by using a harmonic mean of the corresponding sample sizes [8].

After collating the mean answers and the one-way ANOVA statistical test results for each statement, final inferences were made about each global hub in the bank with regards to how their employees communicate and share knowledge in their informal work communities.

4 Results

Summary for Hypothesis 1

Overall, five out of seven questions framed around Hypothesis 1 yields statistically significant results that work to reject the null hypothesis which was there is no difference between the outcomes (effectiveness/efficiency) of knowledge sharing between informal communities in individualist-oriented Geographies and informal communities in collectivist-oriented Geographies. Thus, the answer to majority of the questions has shown that overall, null Hypothesis 1 can be rejected. Here are some key insights from the results:

- Employees working in the Australian hub feel less supported in their job after socialising with a group of colleagues than employees working in the Chinese and Indian hubs.
- Employees working in the Australian hub view informal social settings as a less useful place to share new ideas than employees working in the Indian, Chinese and Philipino hubs.
- Employees working in the Australian hub are less selective with knowledge they share than employees working in Chinese and Philipino hubs.
- Employees working in the Australian hub view the impact of group socialising as less useful to the effectiveness and efficiency of their job in comparison to employees working in the Indian, Chinese and Pilipino hubs.

It is recognised that doing seven tests for Hypothesis 1 makes it more likely one will find significant results, the fact that five out of seven are significant indicates a clear difference between the Australian hub and the rest of the global hubs. As seen in Fig. 1 below, the Australian hub has slightly less mean agreeable answers to the questions in the survey claiming outcomes for work are better for employees socializing in informal communities than other global hubs China has the highest mean agreeable answers to the questions in the survey claiming outcomes for work are better for employees socializing in informal communities. From this, it seems evident that Australia views informal communities as less critical to the successful outcomes (efficiency and effectiveness) of their work than other global hubs.

	AUS	PHL	FJI	IND	CHN	Mean Answer
Q12. After socialising with my work colleagues, I feel more supported/dedicated to my job	4.622	5.157	5.385	5.39	6.174	
Q13. Informal social settings are a great place for me to suggest ideas	4.514	5.154	5.396	5.429	5.913	
Q17. After socialising in informal communties, I learn a more efficient/effective way of doing my job	4.189	4.769	5.086	5.149	5.565	

Fig. 1. Mean answer out of seven of each country for statistically significant questions agreeing with Hypothesis 1

Summary for Hypothesis 2

Overall, three out of seven questions framed around Hypothesis 2 yields statistically significant results that works to reject the null hypothesis which is there is no difference between informal communities in individualist-oriented Geographies and informal communities in collectivist-oriented Geographies with regards to willingness of employees to share their knowledge with other employees. However, four out of seven of the questions framed around Hypothesis 2 yielded results not significant enough to reject null Hypothesis 2. Despite this, some key observations are made from the results that are significant:

- Employees working in the Australian hub are less willing to share information with their colleagues than employees working in the Indian hub.
- Employees working in the Philipino hub are more willing to collaborate with their colleagues in groups than employees working in the Indian hub.
- Employees working in the Australian hub have a lower preference for socialising with their colleagues informally than employees working in the Chinese and Indian hubs.

As seen in Fig. 2 below, it appears China and India have higher preferences for sharing knowledge with colleagues as their overall mean answers to questions in favour of knowledge sharing at the bank appear to be slightly higher than other global hubs. Furthermore, Australia's overall mean answers are slightly lower than the other hubs, with regards to willingness of employees to share knowledge.

	AUS	PHL	FJI	IND	CHN	Mean Answer
Q7. I prefer to work in a collaborative environment	6.419	6.462	6.486	6.667	6.696	
10. I prefer to socialize in a group than be by myself at lunchtime or work hours	4.486	5.2	5.385	5.764	5.826	

Fig. 2. Mean answer out of seven of each country for each statistically significant question agreeing with Hypothesis 2

Summary for Hypothesis 3

Overall, three out of six questions framed around Hypothesis 3 yields statistically significant results that work to reject the null hypothesis which was there is no difference between ICT preferences and frequency of use by employees in informal communities in individualist-oriented Geographies and informal communities in collectivist-oriented Geographies. However, three out of six of the questions framed around Hypothesis 3 yields results not significant enough to reject null Hypothesis 3. Despite this, some key observations are made from the results that are significant:

- Employees working in the Australian hub have a slightly higher preference for communicating through Instant Messenger than employees in the Indian and Philipino hubs.
- Employees working in the Australian hub view the usefulness of ICT tools to express their ideas less favourably than employees in the Indian and Philipino hubs.
- Employees working in the Philipino hub are more prone to using ICT tools as a larger percentage of their workday than employees in the Australian and Indian hubs.

As seen, in Fig. 3, e-mail is by far the most frequently used ICT tool by employees in every global hub.

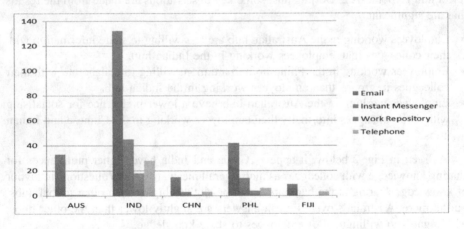

Fig. 3. Frequency of use for ICT tools used for employee communication by country

5 Discussion

In [5, 12, 18], and [10], collectivist-oriented employees will feel the benefits of group interactions more so than individualist-oriented employees. From the study of the Bank, it as found that employees in individualist-oriented geographies such as Australia tend to feel fewer benefits, in terms of the outcomes of their work, from informal communities than employees working in collectivist-oriented geographies like China, India, The Philippines and Fiji. The results in this study are in agreement with the literature.

[10, 18, 25] and [6] assert that collectivist-oriented employees are more willing to share knowledge with their colleagues because they feel a moral involvement with others, and they also have the desire to divulge as much information as they can to help achieve group goals. In this study of the Bank, the general pattern of answers shows that individualist-oriented employees tend to be less willing to share knowledge than collectivist-oriented employees. This somewhat validates [14] claim that employees in collectivist cultures see knowledge as the property of an MNC, whilst those in individualist culture see knowledge as individual property. Some of the findings in this study are in line with literature on culture and the IC dichotomy differences in knowledge transfer in an organisation. These findings somewhat link back to what [12] claims in their study, which is that individualist employees are less likely to divulge information with others in their communities unless it benefits them in some way.

According to [12, 16, 21] and [23], individualist-oriented employees tend to prefer using e-mail and work repositories for communication and collectivist-oriented employees tend to prefer using instant messenger and telephone for communication with their colleagues. In this study of the bank, it was found that e-mail is the most commonly used and preferred ICT tool used by employees across all global hubs, regardless of their cultural orientation. The findings in this study are not in line with literature written on culture, knowledge sharing and ICT preferences.

Based on the findings in this study, recommendations for senior management are given below:

- Senior management leading more individualist-minded employees could make employees feel more positive after every group interaction if they put massive emphasis on the fact that these types of interactions are beneficial to their individual careers.
- Senior management leading more collectivist-minded employees should put in place policies and practices that emphasize how each individual involvement in group settings helps contribute to group goals in the workplace.
- Senior management leading more individualist-minded employees could implement policies and practices that encourages a culture of knowledge transparency between workers for the benefits of their individual careers e.g. a performance-based reward system for sharing knowledge.
- Senior management across all the global hubs should explore various tools that could be used in e-mail to encourage knowledge sharing in an innovative way, since this is an ICT tool already so widely used in both collectivist-oriented and individualist-oriented offices i.e. a blogging feature, e-mail tagging or any add-on software that is user-friendly and encourages knowledge sharing.

6 Conclusion and Further Work

In conclusion, the cultural dimension of "individualism" and "collectivism" does to a degree affect how employees communicate and collaborate in their respective global offices. However, these cultural dimensions seem to have had less of an impact on how employees engage with ICT networking technologies. In this direction, suggestions for further work include doing a study on organisational culture versus geographic culture or doing a study on organisations in transitioning geographic cultures.

As this study is based on a single global bank, the findings may not be generalizable to other types of industry. In particular, for high technology intensive industries where professionalism of an individual may be compromised due to organizational goals in specific cultural based context. Our further work will take the direction of longitudinal study for specific industry type for classification of informal organisation structure versus firm performance.

References

1. Bachmann, D., Elfrink, J., Vazzana, G.: Tracking the progress of e-mail vs. snail-mail. Mark. Res. **8**, 31–35 (1996)
2. Bernard, H.R.: Research Methods in Anthropology: Qualitative and Quantitative Approaches. Rowman Altamira, Lanham (2011)
3. Bhagat, R.S., Ford, D., Jones, C.A., Taylor, R.: Knowledge management in global organizations: implications for international resource management. Res. Pers. Hum. Resour. Manag. **21**, 243–274 (2002)
4. Brunetto, Y., Xerri, M.J., Nelson, S., Farr-Wharton, B.: The role of informal and formal networks: how professionals can be innovative in a constrained fiscal environment. Int. J. Innov. Manag. **20**(3), 1650051-1–1650051-27 (2016)
5. Chen, J., Sun, P.Y., McQueen, R.J.: The impact of national cultures on structured knowledge transfer. J. Knowl. Manag. **14**(2), 228–242 (2010)
6. Chen, G., Tjosvold, D., Li, N., Fu, Y., Liu, D.: Knowledge management in Chinese organizations: collectivist values for open-minded discussions. Int. J. Hum. Resour. Manag. **22**(16), 3393–3412 (2011)
7. De Long, D.W., Fahey, L.: Diagnosing cultural barriers to knowledge management. Acad. Manag. Executive **14**, 113–128 (2000)
8. Elliott, A.C., Woodward, W.A.: IBM SPSS by Example: A Practical Guide to Statistical Data Analysis. SAGE Publications, London (2015)
9. Hofstede, G.: Culture and Organizations: Software of the Mind. McGraw-Hill, London (1991)
10. Hofstede, G.: Culture and Organizations: Software of the Mind. McGraw-Hill Co., New York (1997)
11. Hofstede, G.: Culture's Consequences. SAGE, Beverly Hills (2001)
12. Ismail, K.M.: Theorizing on the role of individualism-collectivism in tacit knowledge transfer between agents in international alliances. Int. J. Knowl. Manag. (IJKM) **8**(1), 71–85 (2012)
13. Janhonen, M., Johanson, J.E.: Role of knowledge conversion and social networks in team performance. Int. J. Inf. Manag. **31**(3), 217–225 (2011)
14. Kedia, B.L., Bhagat, R.S.: Cultural constraints on transfer of technology across nations: implications for research in international and comparative management. Acad. Manag. Rev. **13**(4), 559–571 (1988)
15. Kostova, T.: Success of the transnational transfer of organizational practices within multinational corporations, unpublished doctoral dissertation. University of Minnesota, Minneapolis (1996)

16. Lee, C., Kelkar, R.S.: ICT and knowledge management: perspectives from the SECI model. Electron. Libr. **31**(2), 226–243 (2013)
17. Leung, K.: Negotiation and reward allocations across cultures. In: Earley, P.C., Erez, M. (eds.) New Perspectives on International Industrial/Organizational Psychology, pp. 640–675. Jossey-Bass, San Francisco (1997)
18. Lucas, L.M.: The role of culture on knowledge transfer: the case of the multinational corporation. Learn. Organ. **13**(3), 257–275 (2006)
19. McDermott, R., O'Dell, C.: Overcoming cultural barriers to sharing knowledge. J. Knowl. Manag. **5**(1), 76–85 (2001)
20. Muhammad, N.M.N., Isa, F.M., Kifli, B.C.: Positioning Malaysia as Halal-Hub: integration role of supply chain strategy and halal assurance system. Asian Soc. Sci. **5**(7), 44 (2009)
21. Nonaka, I.: The knowledge-creating company. Harvard Bus. Rev. **69**(6), 96–104 (1991)
22. Nonaka, I., Takeuchi, H.: The Knowledge Creating Company. Oxford University Press, New York (1995)
23. Nonaka, I., Konno, N.: The concept of "ba": building a foundation for knowledge creation. Calif. Manag. Rev. **40**, 40–54 (1998)
24. Punch, K.: Introduction to Social Research: Quantitative and Qualitative Approaches, 3rd edn. SAGE Publications, London (2014)
25. Triandis, H.C.: Individualism-collectivism and personality. J. Pers. **69**(6), 907–924 (2001)
26. Schwartz, S.H.: Universals in the content and structure of values: theoretical advances and empirical tests in 20 countries. In: Zanna, M. (ed.) Advances in Experimental Social Psychology, vol. 25, pp. 1–65. Academic Press, New York (1992)
27. Schwartz, S.H.: Beyond individualism/collectivism: New cultural dimensions of values. In: Kim, U., Triandis, H.C., Kagitcibasi, C., Choi, S.-C., Yoon, G. (eds.) Individualism and Collectivism: Theory, Method, and Applications, pp. 85–119. SAGE, Thousand Oaks (1994)
28. Takeuchi, H., Nonaka, I. (eds.): Hitotsubashi on Knowledge Management. John Wiley & Sons, Singapore (2004)
29. Wellman, B.: An electronic group is virtually a social network. Cult. Internet **4**, 179–205 (1997)
30. Wright, K.B.: Researching internet-based populations: advantages and disadvantages of online survey research, online questionnaire authoring software packages, and web survey services. J. Comput. Mediated Commun. **10**(3) (2005)

Effects of Different Trust on Team Creativity: Taking Knowledge Sharing as a Mediator

Jiangning Wu[✉], Hang Zhao, and Donghua Pan

Institute of Systems Engineering,
Dalian University of Technology, Dalian 116024, China
{jnwu,gyise}@dlut.edu.cn, zhaohang@mail.dlut.edu.cn

Abstract. In the highly dynamic and competitive environment, firms increasingly rely on teams and their members' creativity to survive and succeed. Trust has been found to be the critical antecedent of team creativity, whereas what forms of trust and to what extent the trust can affect team creativity are still unexplored comprehensively. In this respect, the article synthesizes three different types of trust and explores their effect mechanisms on team creativity as well as the role of knowledge sharing in mediating the relationship between trust and team creativity. A survey is conducted to a sample of 61 project teams comprising 409 members from small and medium-sized Chinese firms in Dalian. The findings suggest that there exist different ways for trust to affect team creativity through knowledge sharing, of which calculative trust and institutional trust improve the team creativity by sharing the explicit knowledge, and relational trust ad hoc cognitive trust positively contributes to team creativity through sharing the tacit knowledge.

Keywords: Trust · Team creativity · Knowledge sharing · Calculative trust · Relational trust · Institutional trust

1 Introduction

Dynamic and competitive environment has made firms more and more rely on the creativity to survive and succeed. Team as the basic unit of the firm can generate creative ideas for business innovation due to its flexibility and quick responsive capability. Team creativity can be understood as the production of novel and useful ideas concerning products, services, processes, and procedures by a team of employees working together (Shin and Zhou 2007). Creativity can facilitate team members to fulfill complex tasks and solve problems by using new ideas and ways.

Teamwork emphasizes coordination, sharing information, and participative decision making. Of the various factors that affect the team creative performance, trust has been found to be a critical antecedent (e.g. Barczak et al. 2010). In trustworthy team, teammates prefer to share their skills or expertise as they believe that such behaviors can bring them better returns not only in material aspects but also in spiritual aspects. As a result, team effectiveness as well as team creativity will be enhanced. Although prior research has proved the effect of trust on team creativity, the question "How does trust and its related dimensions affect the team creativity" is still unexplored

© Springer Nature Singapore Pte Ltd. 2016
J. Chen et al. (Eds.): KSS 2016, CCIS 660, pp. 44–56, 2016.
DOI: 10.1007/978-981-10-2857-1_4

comprehensively. This leads us to investigate the relationships between trust types and team creativity. Considering that knowledge sharing within the team can provide more chances to new knowledge creation and novel idea generation, knowledge sharing as a mediator between trust and team creativity is also studied.

This study presents the contribution to trust literature by illustrating the formation of trust in the project team to answer the question "What forms of trust and to what extent the trust can affect the team creativity". In the following two sections, we firstly elaborate on the constructs of trust from a synthetic perspective, and then describe our research model and hypotheses.

2 Trust Within the Team

Trust as a complex, multi-level, and multi-dimensional social psychological phenomenon is defined differently across and within disciplines. One widely accepted definition of trust was given by Mayer et al. (1995): the willingness of a party to be vulnerable to the actions of another party based on the expectation that the other will perform a particular action important to the trustor, irrespective of the ability to monitor or control that other party. The concept trust in the team context is the belief that team members have good intentions and also have confidence in the capability and character of team members.

To model trust within the team, two roles including trustor and trustee and two levels regarding individual and team separately should be considered. At the individual level, the trust generated by individual himself whether trustor or trustee is called as calculus-based trust, which is based on rational choice—characteristic of interactions based upon economic exchange (Rousseau et al. 1998). Such trust emerges when the trustor perceives that the trustee intends to perform an action that is beneficial, and vice versa. With repeated interactions over time between trustor and trustee, relational trust also known as interpersonal trust is built progressively, of which affective trust and cognitive trust are two main forms in team development. Affective trust is the confidence one places in a team member based on one's feelings of caring and concern illustrated by that co-worker. Cognitive trust is based on one's willingness to rely on a team member's expertise and reliability (McAllister 1995). At the team level, institutional trust as a control can influence both calculative trust and relational trust. Institutional factors can act as broad supports for the critical mass of trust that sustains further risk taking and trust behavior (Gulati 1995).

To synthesize three different forms of trust, a dual-level model of trust is designed as shown in Fig. 1, which incorporates both individual-focused trust (referring in particular to within-person and interpersonal trust) and team-focused trust (referring in particular to group trust). This integration provides a more comprehensive view to trust constructs in team settings.

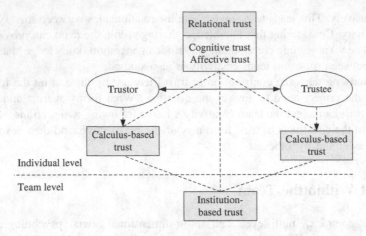

Fig. 1. A model of trust

3   Theoretical Framework and Hypotheses Development

The aim of this research is to discover the influence of trust and knowledge sharing on team creativity. Accordingly a hypothesized model is put forward for this purpose as shown in Fig. 2, in which team creativity is the dependent variable, trust within the team is the independent variable, and knowledge sharing is the mediating variable. The trust variable consists of three constructs: calculative trust, relational trust and institutional trust. For relational trust, only cognitive trust is considered in the study. The reason why to do this is explained in the following section. Moreover, the inter-relations between different forms of trust displayed in Fig. 1 are ignored just for the sake of simplification.

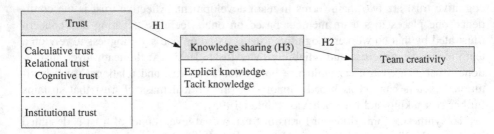

Fig. 2. Hypothesized model

3.1   Trust and Team Creativity

A team is normally understood to be a group whose members are often in charge of complex tasks and are typically cross-functional. Team members in general hold divergent knowledge and skills in different areas, and they may come from different functional departments in the firm. Team is able to fulfill tasks or projects successfully just depending on its ability to integrate the relevant knowledge and skills that are distributed among its members. This integration of the capabilities in the team depends on the way teammates work together and their interpersonal relationships, such as the degree of trust.

Trust within the team has been associated with several outcomes that are expected to contribute positively to the success of task or project. Such outcomes include knowledge sharing, team satisfaction, formation of social networks, team performance, etc. Increased knowledge sharing in the team may lead to a better utilization of existing knowledge and improve the capability for problem solving.

In addition, trust is crucial for the innovation process, due to the risks and uncertainties inherent in creating and implementing novel ideas. Uncertainties and risks may take the form of opportunistic behaviors, failure of technology, unforeseen organizational hurdles and financial concerns. Trust allows team members involved in innovation to share knowledge and collectively solve problems to manage these risks (Shazi et al. 2014).

Numerous scholars have emphasized the importance of trust in the success of innovations. Although the current work on the relationship between trust and team creativity remains largely inconclusive (Bidault and Castello 2009), there is no doubt that trust can inspire both individual and team creativity significantly. Chen et al. (2008) declared that mutual trust has a positive influence on creativity in teams. Although some relationships between trust and creativity have been found in previous studies, we have to go further than current work with respect to dimensions of trust. Therefore, this study focuses on trust types and their roles in teamwork settings where members are highly required to present their creativity in completing their tasks or projects.

As stated above, trust plays an important role in knowledge sharing between team members. Considering three types of trust addressed in Sect. 2, institutional trust has been proven to be imperative for knowledge sharing among team members. Members need to rely on each other's fear of the punishment for not abiding to the rules and regulations of the team as a basis of trust to share knowledge. As one type of relational trust, Holste and Fields (2010) pointed out the cognitive trust is significantly related to knowledge sharing. In the contrast, affective trust based on the emotional bond between trustor and trustee does not result from reasoning and understanding but from feeling and sense, and therefore can be inferred as an indirect factor to knowledge sharing. In this regard, affective trust is eliminated from the proposed model of trust, and the other four types of trust are involved in the further study.

3.2 Trust, Knowledge Sharing and Team Creativity

Trust in teammates may represent a foundation upon which team members feel free to share knowledge, explore, and contribute to the best of their ability to attain successful task completion; this is particularly true when collaboration is required and when creative outcomes are to be achieved (Barczak et al. 2010). Barczak's work also revealed that there is the link between team trust and creative performance. Moreover, the other research found that teams in which trust was high outperformed teams in which trust was low in terms of creative problem solving. In short, trust can be linked with team creative performance. From this point, the hypothesis on the relationship between trust and knowledge sharing can be given as following:

H1. Trust within the team positively impacts on knowledge sharing.

Knowledge sharing is a behavior by which team members exchange and discuss knowledge with internal or external teams through varied ways such as face-to-face discussion, informal and formal networks, and best practices. Knowledge sharing often occurs when team members need to develop and facilitate creativity further by expanding the value of knowledge use (Chae et al. 2015). In knowledge sharing behavior, team members' action is influenced by team-based factors say trust as well as individual-based factors namely perceived relative advantage (Lin et al. 2009). Regarding individual him/herself, once a team member perceives the advantages ad hoc expertise from the others, he/she would like to ask for help to share their knowledge to improve his/her ability to deal with difficult problems. With frequently exchanging thoughts and skills between team members, the whole level of knowledge in the team will be promoted as a result.

Knowledge sharing is not only beneficial to the individual, but also beneficial to the improvement of team performance as well as the promotion of team creativity. Lee et al. (2011) proved by experiments that knowledge sharing is related positively to individual creativity. The reason lies in that the more individuals involve in knowledge sharing and discussion, the more likely they are to contribute new ideas and advanced ways to problem solving. Another fact discovered by Tsai (2000) is that a high level of knowledge sharing is associated positively with innovation. In fact, the interpersonal interaction including knowledge exchange and sharing between team members has significant and positive effects on team creativity (Chen et al. 2008). Based on these arguments, the study advances the following hypothesis:

H2. Knowledge sharing is related positively to team creativity.

Under the hypothesis **H1** mentioned above, we further consider the different forms of knowledge and how they impact creative outcomes since less attention has been focused on these issues. Polanyi (1966) established the two-dimensional concept of explicit and tacit knowledge. Explicit knowledge can be readily articulated, codified, accessed and verbalized. It can be easily transmitted to others. Different from the explicit knowledge, tacit knowledge remains implicit which is difficult to transfer to another person by means of writing it down or verbalizing it. Tacit knowledge includes operational skills and know-how and has roots in actions, procedures or routines. Tacit knowledge is accessible if it is converted to the explicit one. The creation of new knowledge involves the interaction between both modes of knowledge and, thus, the creative performance of a team is critically dependent on both mobilizing tacit knowledge and fostering interaction with explicit knowledge. Therefore, we adopt Sharifirad's (2016) finding which indicates that the sharing of information and know-how within a team by individuals is essential for knowledge creation and is thus the foundation of team's creative performance. In light of two modes of knowledge, hypotheses 1 and 2 can be extended as follows:

H1a. Trust positively impacts explicit knowledge sharing.

H1b. Trust positively impacts tacit knowledge sharing.

H2a. Explicit knowledge sharing is positively related to team creativity.

H2b. Tacit knowledge sharing is positively related to team creativity.

Further, we explore the mediating role of knowledge sharing in the interaction between trust and team members' creative performance. Knowledge sharing is a two-way process where individuals mutually exchange their knowledge and jointly create new knowledge and then enhances their creativity. According to social exchange theory, a good social relationship between two sides of knowledge sharing can be generated when knowledge recipient receives an initial offer of knowledge. If knowledge sender perceives that the recipient reciprocates properly, then trustworthiness between them is confirmed, and exchange relations can be established. Since knowledge can be viewed as a type of asset that cannot be changed by pricing, social exchange theory explains that knowledge sharing can take place only when expected reciprocal benefits between knowledge sender and recipient meet each other's expectations. Based on the social exchange theory, trust is an important factor affecting knowledge sharing. In other words, trust is the basic requirement for knowledge sharing, which implies the mediating role of knowledge sharing between trust and team creativity. Taken together, this leads us to the following hypothesis:

H3. Knowledge sharing plays an intermediating role between trust and team creativity.

In terms of three forms of trust and two modes of knowledge, specifically the hypothesis 3 can be extended to the following sub-hypotheses:

H3a1. Explicit knowledge sharing plays an intermediating role between calculus-based trust and team creativity.

H3a2. Explicit knowledge sharing plays an intermediating role between cognitive trust and team creativity.

H3a3. Explicit knowledge sharing plays an intermediating role between relational trust and team creativity.

H3a4. Explicit knowledge sharing plays an intermediating role between institution-based trust and team creativity.

H3b1. Tacit knowledge sharing plays an intermediating role between calculus-based trust and team creativity.

H3b2. Tacit knowledge sharing plays an intermediating role between cognitive trust and team creativity.

H3b3. Tacit knowledge sharing plays an intermediating role between relational trust and team creativity.

H3b4. Tacit knowledge sharing plays a intermediating role between institution-based trust and team creativity.

4 Methods

4.1 Data

Data used to test the research model were collected from Chinese small and medium-sized firms in Dalian city. Questionnaires were passed to individuals who are working in different project teams from different firms. The questionnaire required the respondents to answer the questions regarding four different types of trust (i.e.

calculus-based trust, relation-based trust, cognition-based trust and institution-based trust), their knowledge sharing behavior, and their evaluation of the effectiveness of teams that they are involved with. The survey delivered 487 questionnaires from April 2015 to July 2015, and 409 were returned to the end with 83.98 % return rate. In total, 409 individuals in 61 project teams were invited to participate in this empirical study. The demographic profile for the respondents indicated that 65.5 % were male and 34.5 % were female. Further, more than half of the sample (53.1 %) was between 26 to 30 years old. 97 % of the respondents had bachelor or even higher degrees. The average team tenure was 3.36 years (SD = 2.1). As for team size, most of the sample (37.2 %) holds 6 to 10 members.

4.2 Measures

A total 38 items were used to measure three factors: (1) trust, (2) knowledge sharing, and (3) team creativity. Respondents made responses to all the items on a seven-point Likert-type scale (1 = strongly disagree; 7 = strongly agree).

Trust

Trust was measured by a 24-item scale, which is on the basis of mature abroad research scales and domestic research results. We measured calculus-based trust by six items, which drew lessons from Butler (1991). A sample item was "If team members are committed to me, I feel that they could keep their promises". Cognition-based trust was measured by six items, which adopted by McAllister (1995). A sample item was "I suppose team members show a high level of professional skills and knowledge in the work". For the above two scales, Cronbach's α = 0.85. Next, six items were adopted by both Wong et al. (2003) and Chu (2008) to measure relation-based trust (Cronbach's α = 0.86). A sample item was "I feel that team members put much affection in keeping interpersonal relationship". And six items adopted by Moorman et al. (1993) were used to measure institution-based trust (Cronbach's α = 0.84). A sample item was "There are some rules and regulations within the team, and the team members have to abide by them". All the designed items show satisfactory internal consistency.

Knowledge Sharing

Knowledge sharing measures include explicit knowledge sharing and tacit knowledge sharing. These two variables were measured by four items based on both Bock and Kim (2001) and Bock et al. (2005), respectively. An example of explicit knowledge sharing measured item was "I will provide team members with the tools I use to work with". And a sample item of tacit knowledge sharing measured item was "If team members need my assistance, I would like to provide my own understanding of the source of knowledge or clues to the insiders". The Cronbach's alpha coefficients of two scales were 0.84 and 0.83, respectively.

Team Creativity

Team creativity was measured by six items taken from Amabile et al. (1996), focusing on team level output. An example of item was "My team often has a creative solution to the current problems". Cronbach's alpha of aggregated scores was 0.9, showing a high internal consistency.

5 Results

A two-step approach to structural equation modeling (SEM) was applied in the empirical study. Confirmatory factor analysis (CFA) was conducted to assess the proposed measurement model fit and construct validity. The structural model was generated to test the significance of the theoretical relationships.

5.1 Confirmatory Factor Analysis of Trust Scale

In order to test the model fitting of trust scale, a CFA of 204 valid data was conducted using AMOS 21.0 to verify the results of exploratory factor analysis from different perspectives. Detailed results are shown in Fig. 3.

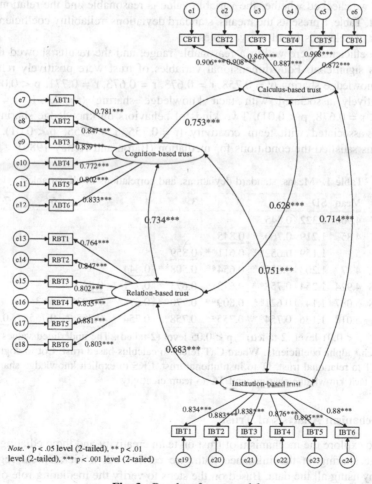

Note. * p < .05 level (2-tailed), ** p < .01
level (2-tailed), *** p < .001 level (2-tailed)

Fig. 3. Results of trust model

The factors of the four potential variables of trust to each measurement index were ranging from 0.5 to 0.95, which means the basic fitness of the trust model was good. All the goodness of fit indices for the four-factor model respectively were: $\chi^2/df =$ 1.798 (reference values 1.0–3.0), RMSEA (Root Mean Square Error of Approximation) = 0.063 (< 0.08 reference value), GFI (Goodness-of-Fit Index) = 0.847 (> 0.8 reference value), AGFI (Adjusted Goodness-of-Fit Index) = 0.813 (> 0.8 reference value), NFI (Normed Fit Index) = 0.918 (> 0.9 reference value) and IFI (Incremental Fit Index) = 0.962 (> 0.9 reference value). According to the above indicators, the fitness of the trust model was well.

5.2   Correlation Analysis of Variables

To test the reliability of each variable, we conducted an analysis of all data using SPSS22.0. At the same time, by means of descriptive statistics and correlation measurement, we checked whether the variable value is reasonable and the relationship is significant. Table 1 presents the means, standard deviations, reliability coefficients, and zero-order correlations of all the variables.

The coefficients were within a reasonable range, and the results showed that the effect was significant. Four-dimensional variables of trust were positively related to explicit knowledge sharing (r = 0.758, r = 0.673, r = 0.673, r = 0.771, p < 0.01), and also positively associated with tacit knowledge sharing (r = 0.621, r = 0.809, r = 0.770, r = 0.678, p < 0.01). Two kinds of behaviors on knowledge sharing were positively associated with team creativity (r = 0.755, r = 0.755, p < 0.01). These correlations satisfied the conditions for mediation (Baron and Kenny 1986).

Table 1. Means, standard deviations, and correlations of all variables

	Mean	SD	1	2	3	4	5	6	7
1. CaT	4.906	1.322	0.845						
2. CoT	4.85	1.219	0.760**	0.845					
3. RT	5.1	1.159	0.652**	0.611**	0.859				
4. IT	4.87	1.203	0.630**	0.654**	0.508**	0.844			
5. EKS	4.924	1.254	0.758**	0.673**	0.673**	0.771**	0.84		
6. TKS	5.087	1.14	0.621**	0.809**	0.770**	0.678**	0.693**	0.83	
7. TC	5.018	1.136	0.754**	0.755**	0.758**	0.754**	0.755**	0.755**	0.901

Note. ** p < 0.01 level (2-tailed) * p < 0.05 level (2-tailed). The underlined values are Cronbach's alpha coefficients. Where CaT refers to calculus-based trust, Cot to cognitive trust, RT to relational trust, IT to institutional trust, EKS to explicit knowledge sharing, TKS to tacit knowledge sharing, and TC to team creativity.

5.3   Mechanism Analysis of Trust

In order to explore the mechanism of trust on team creativity and the mediating role of knowledge sharing in the influence path, we carried out a hierarchical regression analysis by using all the data. Based on the steps to verify the mediating role of Baron and Kenny (1986), we constructed five regression models. Model I: Exploring the impact of trust on team creativity, which includes calculative trust, cognitive trust, relational trust, and institutional trust. Model II: Investigating the effect of team trust on explicit knowledge sharing. Model III: Exploring the joint effects of trust and explicit knowledge sharing on team creativity to test the mediating role of explicit knowledge sharing. Model IV: Investigating the impact of trust on tacit knowledge sharing. Model V: Exploring the joint effects of trust and tacit knowledge sharing on team creativity, which is used to check the intermediation of tacit knowledge sharing. Table 2 below shows the values of each detailed indices.

From the data in Table 2, it can be seen that the beta values representing the effect of trust on team creativity have been changed after the introduction of knowledge sharing intermediary factors. Focusing on models III and V, it can be found that the beta values decreased not to a significant level, which indicated that the intermediary factors played a full intermediary role in this case. Therefore the corresponding hypotheses H3a1, H3a4, H3b2, and H3b3 were supported, which "Explicit knowledge sharing plays an intermediating role between calculus-based trust and team creativity", "Explicit knowledge sharing plays an intermediating role between institution-based trust and team creativity", "Tacit knowledge sharing plays an intermediating role between cognitive trust and team creativity", and "Tacit knowledge sharing plays an intermediating role between relational trust and team creativity" were verified by analysis. When the beta coefficients decreased but they were still at a significant level which indicated that the intermediating factors played a part of intermediary role. Thus hypotheses H3a2, H3a3, and H3b4 were partially verified, which "Explicit knowledge sharing plays an intermediating role between cognitive trust and team creativity", "Explicit knowledge sharing plays an intermediating role between relational trust and team creativity", and "Tacit knowledge sharing plays a intermediating role between institution-based trust and team creativity" were partially verified. In the experiment, the hypothesis H3b1 that "Tacit knowledge sharing plays an intermediating role between calculus-based trust and team creativity" had not been verified.

Table 2. Results of regression analysis of all models

	Model I	Model II	Model III	Model IV	Model V
	beta	beta	beta	beta	beta
Dependent variables	TC	EKS	TC	TKS	TC
Independent variables					
CaT	0.155*	0.339***	0.069	0.228***	0.167**
CoTt	0.274***	0.212***	0.221***	0.361***	0.101
RTt	0.233***	0.165*	0.192**	0.128*	0.124
IT	0.213***	0.187**	0.115	0.199***	0.189**
Intermediate variables to be verified					
EKS			0.234***		
TKS					0.252***
F-value	167.137***	222.854***	138.132***	245.945***	135.299***
	0.62	0.685	0.627	0.706	0.622

Note. *** $p < 0.001$ level (2-tailed) ** $p < 0.01$ level (2-tailed) * $p < 0.05$ level (2-tailed).

6 Discussions and Conclusions

This study aims to provide the insight regarding the relationship between different trust and team creativity by mediating knowledge sharing. Although trust research in organization/team science has adopted an agreed definition and multidimensional traits of trust, different types of trust and their relationships to team creativity have been less investigated, particularly in project team management, which is the main research focus in this study.

By constructing a dual-level model of trust model and conducting an empirical study, three meaningful findings have been revealed.

First, calculative trust positively affects team creativity through explicit knowledge sharing, whereas tacit knowledge sharing makes no sense. This finding is true in the condition of team member being rational. It can be explained that during face-to-face communications, team members with certain trust relationships no matter trustor or trustee initially make rational choice to help with each other or not based on the social exchange theory. Once the intention or interest of one side of trustor-trustee is satisfied, he/she is willing to contribute his/her ideas, opinions and information which refer to explicit knowledge. Just because of the rational characteristics, team members are not able to fully involved in interactive activities related to team tasks to protect their own competence which refers to tacit knowledge.

Second, institutional trust has positive effect on team creativity through explicit knowledge sharing. As the team-level trust, the role of institutional trust is often used to advise individuals to obey team's codes or norms. Such restraint in many cases can only work well on individual's behavior but not on his/her mental activity because rules of team may not be perceived as beneficial directly. This leads to the fact that institutional trust plays more important role in explicit knowledge sharing than in tacit knowledge sharing.

Third, relational trust and cognitive trust are positively associated with team creativity through tacit knowledge sharing. This finding implies that deep communication and understanding contributes more to team creativity. With repeated interactions, teammates know well with each other not only in aspects of personal traits like personality but also individual's ability, expertise, etc. The higher level of relational trust and cognitive trust are reached, the more tacit knowledge may be exchanged and shared, which begets a higher team creativity performance to the end.

Based on the main findings mentioned above, we can conclude that different trust needs to be considered together with knowledge interactions in order to achieve team goals more effectively by maximizing each member's creativity.

Although this study expands our knowledge about the effect of different types of trust on team creativity by introducing knowledge sharing as a mediator, some limitations still exist. One limitation lies in the selected samples which are only collected from the same place. Accordingly, the regional phenomenon may be encountered. The other limitation comes from the data itself which is part of the cross section data. Such data can only be used to find correlations between variables. In the follow-up study, the longitudinal data may be added to discover the causality between variables.

Acknowledgements. The work is fully sponsored by the projects of Natural Science Foundation of China (NSFC) under Grant Nos. 71271036 and 71471028.

References

Amabile, T.M., Conti, R., Coon, H., Lazenby, J., Herron, M.: Assessing the work environment for creativity. Acad. Manag. J. **39**(5), 1154–1184 (1996)

Barczak, G., Lassk, F., Mulki, J.: Antecedents of team creativity: an examination of team emotional intelligence, team trust and collaborative culture. Creativity Innov. Manag. **19**(4), 332–345 (2010)

Baron, R.M., Kenny, D.A.: The moderator–mediator variable distinction in social psychological research: conceptual, strategic, and statistical considerations. J. Pers. Soc. Psychol. **51**(6), 1173–1182 (1986)

Bidault, F., Castello, A.: Trust and creativity: understanding the role of trust in creativity-oriented joint development. R&D Manag. **39**(3), 259–270 (2009)

Bock, G.W., Kim, Y.G.: Breaking the myths of rewards: an exploratory study of attitudes about knowledge sharing. In: PACIS 2001 Proceedings, p. 78 (2001)

Bock, G.W., Zmud, R.W., Kim, Y.G., Lee, J.N.: Behavioral intention formation in knowledge sharing: examining the roles of extrinsic motivators, social-psychological forces, and organizational climate. MIS Q. **29**, 87–111 (2005)

Butler, J.K.: Toward understanding and measuring conditions of trust: evolution of a conditions of trust inventory. J. Manag. **17**(3), 643–663 (1991)

Chae, S., Seo, Y., Lee, K.C.: Effects of task complexity on individual creativity through knowledge interaction: a comparison of temporary and permanent teams. Comput. Hum. Behav. **42**(4), 138–148 (2015)

Chen, M.H., Chang, Y.C., Hung, S.C.: Social capital and creativity in R&D project teams. R&D Manag. **38**(1), 21–34 (2008)

Chu, H.N.: The research of enterprise interpersonal trust and its impact on knowledge sharing in Chinese culture environment. Doctoral Dissertation, Huazhong University of Science and Technology (2008). (in Chinese)

Gulati, R.: Does familiarity breed trust? The implications of repeated ties for contractual choice in alliances. Acad. Manag. J. **38**, 85–112 (1995)

Holste, J.S., Fields, D.: Trust and tacit knowledge sharing and use. J. Knowl. Manag. **14**(1), 128–140 (2010)

Lee, D.S., Seo, Y.W., Lee, K.C.: Individual and team differences in self-reported creativity by shared leadership and individual knowledge in an e-learning environment. Information **14**(9), 2931–2946 (2011)

Lin, M.J.J., Hung, S.W., Chen, C.J.: Fostering the determinants of knowledge sharing in professional virtual communities. Comput. Hum. Behav. **25**(4), 929–939 (2009)

McAllister, D.J.: Affect-and cognition-based trust as foundations for interpersonal cooperation in organizations. Acad. Manag. J. **38**(1), 24–59 (1995)

Mayer, R.C., Davis, J.H., Schoorman, F.D.: An integrative model of organizational trust. Acad. Manag. Rev. **20**(3), 709–734 (1995)

Moorman, C., Deshpande, R., Zaltman, G.: Factors affecting trust in market research relationships. J. Mark. **57**, 81–101 (1993)

Polanyi, M.: The Tacit Dimension. Routledge, London (1966)

Tsai, W.: Social capital, strategic relatedness and the formation of intra-organizational linkages. Strateg. Manag. J. **21**(9), 925–939 (2000)

Rousseau, D.M., Sitkin, S.B., Burt, R.S., Camerer, C.: Not so different after all: a cross-discipline view of trust. Acad. Manag. Rev. **23**(3), 393–404 (1998)

Sharifirad, M.S.: Can incivility impair team's creative performance through paralyzing employee's knowledge sharing? A multi-level approach. Leadersh. Organ. Dev. J. **37**(2), 200–225 (2016)

Shazi, R., Gillespie, N., Steen, J.: Trust as a predictor of innovation network ties in project teams. Int. J. Proj. Manag. **33**(1), 81–91 (2014)

Shin, S.J., Zhou, J.: When is educational specialization heterogeneity related to creativity in research and development teams? Transformational leadership as a moderator. J. Appl. Psychol. **92**, 1709–1721 (2007)

Wong, Y.T., Ngo, H.Y., Wong, C.S.: Antecedents and outcomes of employees' trust in Chinese joint ventures. Asia Pac. J. Manag. **20**(4), 481–499 (2003)

Identifying Lead User in Mass Collaborative Innovation Community: Based on Knowledge Supernetwork

Zhihong Li and Hongting Tang[✉]

School of Business Administration, South China University of Technology,
Guangzhou, China
bmzhhli@scut.edu.cn, bmhttang@mail.scut.edu.cn

Abstract. Lead users with advanced demand and innovation capability are those most valuable resources for product development and service innovation. Despite acting as the significant communication platform, the pervasive use of the Internet brings in new challenges to identify valuable users. To identify and analyze those lead users and related knowledge, this paper establishes a knowledge supernetwork model by describing the heterogeneity of agent in mass collaborative innovation community, following an example of a typical community. Based on supernetwork theory, this model has important theoretical implications in the integration of supernetwork method and knowledge management. This study also contributes to providing an insight in recognizing lead users with a visual identification method.

Keywords: Lead user · Knowledge supernetwork · Mass collaborative innovation community · Knowledge management

1 Introduction

With the arising of Web 2.0, virtual community have already penetrated into our daily life [1, 2]. Many large corporations, such as Adobe, Google, Microsoft, have begun to officially build online Mass Collaborative Innovation Community (MCIC) to absorb users' attention and encourage mass users' participation in the process of its product or service innovation. They ultimately commit to realize the collaborative innovation by improving information sharing through those Internet platforms [3].

A great deal of studies pointed out that online users, especially those who are the enterprise's consumers or potential consumers with advanced demand and innovation capability, can designate the corporation's direction to product and service innovation. For example, [39, 40] found that user innovation behavior was widespread in online innovation communities on user innovation communities. Moreover, most users in community developed new products and shared their innovation sources and results for free [4, 5]. Therefore, discerning and effectively utilizing a wealth of knowledge and experience in online innovation community are the key points to gain innovation competitive advantage for enterprises and community managers.

© Springer Nature Singapore Pte Ltd. 2016
J. Chen et al. (Eds.): KSS 2016, CCIS 660, pp. 57–67, 2016.
DOI: 10.1007/978-981-10-2857-1_5

When studying the recognition method of lead user in online community, existing studies have begun to consider user behavior, user labels, and interactive information based on traditional indexes [6–8]. However, there is a lack of efficient expression and analysis method for characterizing the total knowledge resources in innovation community, which is the indefinable and anfractuous relationships among users, posts and knowledge in community. To address such issue, this research thus paves the way for supplement of the researches in this direction.

Based on supernetwork theories and methods, this paper attempts to discover the valuable users and their knowledge in MCIC with the multilayer, multilevel, multi-attribute, and multi-objective analysis. In the empirical validation, we also take full account of the users' attention and the quality of user generated content. The proposed knowledge supernetwork model in this study can be used to help enterprises or community managers to reach the valuable users and related knowledge from large, fragmented, and heterogeneous community.

2 Literature Review

2.1 The Status and Progress of MCIC

Current literature on knowledge management in online community mainly involves driving factors of knowledge innovation [9, 10] and impact factors of knowledge sharing [11–13], especially the motivation of user's knowledge sharing behavior including a large sum of theoretical and empirical analysis in Ask-Answer community [14–16]. However, there is a lack of studies on discovery and analytical methods of lead users and their knowledge.

In addition, as the big data era is coming and the development of network technology, the construction of knowledge finding system and model start to draw many scholars' attentions. Martinez-Torres et al. put forward a tool based on semantic analysis to implement automatic knowledge cognition and text classification in community [17], Liao Xiao et al. constructed a knowledge finding and analysis model based on weighted knowledge network and quantitatively analyzed the users' innovation knowledge in community [18]. Despite their considerable merits, hardly no analytical frameworks achieve the success in expressing users and knowledge elements in the complex community systems with multi-agent and various relationships.

2.2 Knowledge Supernetwork

As the WS small-world network model [19] and BA scale-free network model [20] were proposed, the study on complex network [21] is on the rise around the world. The research interests include social network [22, 23], knowledge network [24], technology network [25, 26] and so on. But when it comes to multiple network interweaved with each other in the study of a massive scale network, the methods mentioned above are not strict enough for clarifying the complex relationships. Therefore, some scholars have attempted to use the supernetwork to express the multi-nodes complex network system, and have made fruitful achievements [27].

On the field of knowledge management, Seufert et al. [28] taken knowledge network as a dynamic framework which constituted by actors, relationships among actors and corresponding institutions and resources. Based on studies on the knowledge management, Xi Yunjiang [29–31] proposed that knowledge network can be classified into knowledge network, member network and material carrier network of knowledge, and he brought forward the weighted knowledge network (WKN) model, weighted knowledge supernetwork (WKSN) model. Compared with existing study results, this paper introduces an offline supernetwork method into online MCIC and builds an online knowledge supernetwork model.

2.3 Lead User and Identification Method

Lead user is used to describe individuals with extensive knowledge and ideas about product improvement and innovation [7]. The current study of lead user is concentrated on three aspects: role and value of lead user [32, 33], innovation motivation and influence of leader [34, 35], and the recognition method of lead user, which contains two main traditional methods, pyramid method and index method (by questionnaire).

With the development of Internet technology, users are more willing to contribute the product knowledge and use experiences, share the product demands and preferences in online community [36, 37]. As a district of product consumers, innovation community is becoming an important research object of lead user. The relevant study about the discovery of lead user online still emphasizes on questionnaire, while adds user behavior, user label, and interactive behavior based on original index system. Thus it can be seen that the recognition method is full of subjectivity. Moreover, the huge quantity of data in community leads the identification of user information to be extremely difficult and inefficient. In this case, the objective of this study is to understand how to express the user and knowledge resources with huge system and heterogeneity. How to define the primacy of user in MCIC based on big data and how to construct effective assessment criteria in Internet environment? The sequencing issues are worthy of exploring further.

3 The Model and Example

3.1 Knowledge Supernetwork Model in MCIC

In MCIC, user's tacit knowledge can be externalized into explicit knowledge by posting behavior, and user's knowledge interactions are embodied by browsing and replying behavior. As a key element in MCIC, knowledge is addressed and involved by posts. The correlation of multi-agent in community can be showed in Fig. 1.

The knowledge supernetwork model contains three layers: K-K knowledge network, U-U user network and P-P post network. According to the supernetwork theory and method, this model can be represented efficiently as shown in Eq. (1).

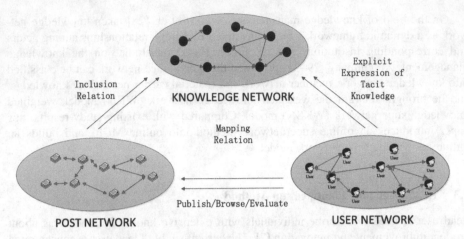

Fig. 1. Conceptual diagram of elements in MCIC

$$G = \left\{ G_u, G_p, G_k, E_{U-P}, E_{P-K}, E_{U-K} \right\} \tag{1}$$

Where $G_u = (U, E_u)$ is the co-occurrence network among community of users, $U = \{u_1, u_2, \cdots, u_m\}$ denotes the set of knowledge owners, m is the number of users, and $E_u = \{\{(u_i, u_j)|\theta(u_i, u_j) = 1\}\}$ is the set of edges that denotes co-occurrence relationships in same post between users.

$G_p = (P, Q_p, E_p)$ is the similarity network among posts, $P = \{p_1, p_2, \cdots, p_n\}$ denotes the set of posts, n is the number of posts, and $Q_P = \{q_{p_1}, q_{p_2}, \cdots, q_{p_n}\}$ is the set of the user attention which depends on views and comments of each post, $E_p = \{\{(p_i, p_j)|w(p_i, p_j) \geq w_0\}\}$ is the set of edges that denotes similar value determined by its contents between two posts is greater than the threshold w_0.

$G_k = (K, Q_K, E_k, W(E_k))$ is the co-occurrence network among knowledge points, $K = \{k_1, k_2, \cdots, k_r\}$ denotes the set of knowledge points, $Q_K = \{q_{k_1}, q_{k_2}, \cdots, q_{k_r}\}$ is the set of the stock which depends on its frequency and position, $E_k = \{\{(k_i, k_j)|w(k_i, k_j) \geq w'_0\}\}$ is the set of edges that denotes the co-occurrence frequency in same post between points is greater than the threshold w'_0 and $W(E_k) = \{w(k_i, k_j)|w(k_i, k_j) \geq w'_0, k_i, k_j \in K\}$ is the set of concrete values of co-occurrence frequency.

Moreover, there are three kinds of relationship mapping between each two agents, E_{U-P} denotes the user post updates, E_{P-K} denotes the knowledge points is contained by the posts and E_{U-K} denotes the user mastered the knowledge points. And three kinds of subnetworks mentioned above are interconnected by those relationship mappings to generate a knowledge supernetwork.

3.2   The Establishment Process of Knowledge Supernetwork Model

The knowledge supernetwork model can be constructed as above in Fig. 2.

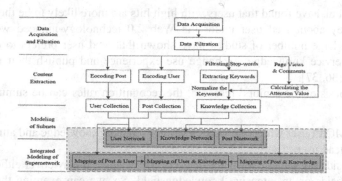

Fig. 2. The modeling process of knowledge supernetwork model

3.3   Example from MIUI Community

In this paper a typical MCIC-MIUI Community is selected as our research object. We collected 23929 posts published before November 13, 2015 in the 'Hongmi Note 2' section by LocoySpider V9.

According to the above method, a knowledge supernetwork model has been constructed by multi-nodes and multiple edges, as shown in the Fig. 3 above. There are three hierarchies. For example, red dots represent users, green dots represent posts and purple dots represent knowledge points, and the size of those circles denotes the node degree.

Fig. 3. Knowledge supernetwork model of MCIC (Color figure online)

4 Recognition Method of Lead User

4.1 Recognition Rules

C.L. Hung et al. have found that users with high hits are more likely to be the lead user after a survey about lead user under the Web 2.0 technology-induced world [38]. Moreover, a large number of studies have shown that lead user prefers to contribute product or service knowledge, exchange use experience and publish their needs and preferences [36, 37].

On the basis of the seniors' research, the recognition rules can be summarized as following:

Rule 1. In MCIC, user who contributes a whole wealth of knowledge and attracts more attention is lead user.

And considering the circumstance of users' knowledge, it may cover either most of knowledge fields or an intensive knowledge field. So, we can examine the users in different fields, such as core innovation field with high attention value and high importance to the development of enterprise, to find specific lead users.

Rule 2. In MCIC, user who contributes a whole wealth of knowledge in core innovation field and attracts more attention among posts in this field is lead user of core innovation field.

4.2 Recognition of Lead User

User's online behavior has been the reflection of his lead in community, thus we can calculate the knowledge stocks of each posts by mapping relation between posts and knowledge, and then calculate the knowledge quantity and attention value of each user to obtain knowledge set $Q_K(U)$ and attention set $Q_P(U)$ respectively. Afterwards the user whose knowledge quantity $Q_K(u_i)$ exceeds the threshold w_1 and attention value $Q_P(u_i)$ exceeds the threshold w_2 will be selected as a lead user. Namely, the collection of lead user can be expressed as $U_L = \{u_i | Q_K(u_i) > w_1, Q_P(u_i) > w_2\}$ and the threshold values are based on business real requirements.

In this case, we selected top-100 users as our lead users as the Table 1 shows. In addition, we reveal the lead users' knowledge supernetwork by intercepting the corresponding subgraph, which is shown in Fig. 4.

4.3 Recognition of Lead User in Core Innovation Field

Core innovation field is crucial to the future development of corporation as a field concerned commonly by user, and it determines the innovation directions and key points of product or service. Before the recognition of lead user in core innovation field, we give out a delicate scope of core innovation field, which is determined by attention value of knowledge $Q_P(k_i)$ based on mapping relation between posts and knowledge. We select the knowledge whose attention value is greater than threshold w_3 (depends on convergence requirement of company) as core innovation field. And we

Table 1. The leaderboard of lead user in community

User no.	User no.	User no.	User no.	User no.	User no.	User no.	User no.	User no.	User no.
U2	U25	U321	U157	U159	U544	U350	U3767	U10	U307
U1	U47	U38	U696	U2039	U62	U187	U242	U330	U245
U33	U127	U5	U3747	U96	U657	U78	U130	U160	U3337
U56	U228	U20	U720	U338	U3172	U1363	U28	U169	U12
U37	U1069	U315	U156	U70	U1898	U112	U206	U1104	U180
U13	U9	U32	U553	U773	U110	U77	U152	U1199	U775
U68	U45	U59	U16	U1382	U60	U3404	U15	U168	U31
U36	U154	U93	U75	U101	U493	U530	U24	U718	U19
U3	U484	U23	U14	U740	U2981	U3750	U757	U34	U2355
U21	U8	U17	U1507	U1814	U39	U3757	U18	U50	U71

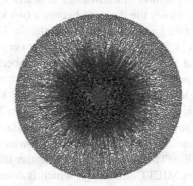

Fig. 4. Knowledge supernetwork model of Top-100 lead user in community (Color figure online)

can obtain the knowledge quantity set $Q'_K(U)$ and attention set $Q'_P(U)$ refer to the step above, then selecting the user whose knowledge quantity $Q'_K(u_i)$ exceeds the threshold w'_1 and attention value $Q'_P(u_i)$ exceeds the threshold w'_2 as a lead user in core innovation field. Namely, the collection of lead user in core innovation field can be expressed as $U'_L = \{u_i | Q'_K(u_i) > w'_1, Q_P(u_i) > w'_2\}$.

Particularly, in the process of discovery of lead user, we can focus on users in a certain knowledge field or users with a specific property in accordance with actual requirements besides users in community or core innovation field.

4.4   Individual Supernetwork Analysis of Lead User

As an effective expression tool of integrated elements, besides the recognition of lead user, we can analyze a certain lead user's knowledge or characteristics by his individual knowledge supernetwork. In this paper we take the lead user 'U2' as an example and analyze his knowledge compositions and structures.

Figure 5 shown a subnetwork of user U2 from knowledge supernetwork. To represent the node property more clearly, we rearrange the layout of the subnet by Gephi. By statistics, there are 730 innovation knowledge points contributed by U2 represented with purple dots, 57 posts published by U2 represented with green dots, and 537 users sharing same interest with U2 represented by red dots. As can be seen from the illustration, both the corresponding users and knowledge take up one second of the layout, and the related posts distributed in the outside of knowledge points. In fact, the user is more important as the user near the border of knowledge and user, and the knowledge stock is larger as the knowledge point near the border of knowledge and users. Likewise, the post's position gets closer to the center of network, the knowledge stock and user attention are higher. So, we can obtain the users matched to user U2, knowledge field mastered by user U2 or corresponding high-quality posts.

To further clarify the knowledge system of user U2, the co-occurrence relationship among knowledge are represented visually by Fig. 6. As the figure shows, the contribution stock is higher, the position of knowledge is nearer to center of network, and vice versa. Besides that, the closer the distance between two knowledge on the graph, the higher the relevance of two knowledge, for example, the knowledge points Note2, Apple(苹果) and plagiarize(抄袭) in a central location are very close to each other, that is because user U2 posts that XiaoMi's partial design is cribbed from Apple. Moreover, the knowledge along the periphery of the network does have obvious aggregation relationship, such as the left edges in the network, the knowledge points resistance(阻力), surface(表面), experience(经验), friction(摩擦), antiskid(防滑), blue(蓝色), sticker(贴纸), rough(粗糙), touch(接触), satisfaction(满意) and van der Waals(范德华) are much closer together than other knowledge points, and all those points are associated with film. Thus it can be seen that user U2 recommends the cover sticker of XiaoMi mobile in MIUI Community, which is consistent with the practical data. Therefore, we can further exploit the potential of user's knowledge to refine users' knowledge situation.

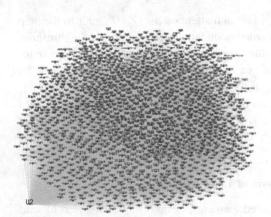

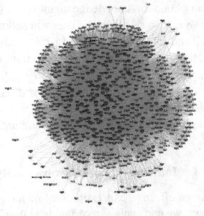

Fig. 5. Knowledge supernetwork model of user U2 (Color figure online)

Fig. 6. Weighted knowledge network of user U2 (Color figure online)

5 Conclusion

On the basis of summary, the paper addresses the research gap of existing research and explores the discovery and analytical method of lead user and his/her knowledge in mass collaborate innovation community. Based on the supernetwork theories, we construct a knowledge supernetwork model to satisfy the requirement of heterogeneous expression of multi-agent and the recognition of lead user in MCIC. At the same time, this article cites the actual data from a typical MCIC - MIUI Community to validate this model. For one thing, we attempt to recognize the lead user in community and core innovation field. For another, we turn to analyze the individual knowledge structure of lead user by corresponding subgraph of knowledge supernetwork.

According to the empirical validation, it can be seen that the discovery method provides a kind of deeper, clearer and visual tool to describe the multi-agent and multi-relation characteristics of knowledge sources in MCIC. Meanwhile, the constructed knowledge supernetwork model promotes the fusion of supernetwork and knowledge management and is of great importance for development of supernetwork. Nevertheless, this study is limited to a relatively low accuracy measurement and choice of knowledge due to the subjectivity and ambiguity of the knowledge, which obviously require further investigation.

Acknowledgements. This research is supported by Major Program of National Science Foundation of China (Project No.71090403/71090400), Program for Natural Science Foundation of Guangdong Province, China (2014A030313243), and China Postdoctoral Science Foundation (2016T90788, 2015M582389).

References

1. Simmons, G.: Marketing to postmodern consumers: introducing the internet chameleon. Eur. J. Mark. **42**(3/4), 299–310 (2008)
2. Jayanti, R.K., Singh, J.: Pragmatic learning theory: an inquiry-action framework for distributed consumer learning in online communities. J. Consum. Res. **36**(6), 1058–1081 (2010)
3. Gloor, P.A.: Swarm Creativity: Competitive Advantage Through Collaborative Innovation Networks. Oxford University Press, Oxford (2005)
4. Von Krogh, G., Von Hippel, E.: The promise of research on open source software. Manag. Sci. **52**(7), 975–983 (2006)
5. Nambisan, S.: Designing virtual customer environments for new product development: toward a theory. Acad. Manag. Rev. **27**(3), 392–413 (2002)
6. Yang, B., Liu, W.: Research on lead user identification on the basis of application extending and netnews. Chin. J. Manag. **08**(9), 1353–1358 (2011)
7. Von Hippel, E.: Lead users: a source of novel product concepts. Manag. Sci. **32**(7), 791–805 (1986)
8. Morrison, P.D., Roberts, J.H., Midgley, D.F.: The nature of lead users and measurement of leading edge status. Res. Policy **33**(2), 351–362 (2004)

9. Sheng, M., Hartono, R.: An exploratory study of knowledge creation and sharing in online community: a social capital perspective. Total Qual. Manag. Bus. Excellence **26**(1–2), 93–107 (2015)

10. Lou, J., et al.: Contributing high quantity and quality knowledge to online Q&A communities. J. Am. Soc. Inf. Sci. Technol. **64**(2), 356–371 (2013)

11. Park, J.H., et al.: An investigation of information sharing and seeking behaviors in online investment communities. Comput. Hum. Behav. **31**, 1–12 (2014)

12. Tsai, H.-T., Bagozzi, R.P.: Contribution behavior in virtual communities: cognitive, emotional, and social influences. MIS Q. **38**(1), 143–163 (2014)

13. Shan, S., et al.: Identifying influential factors of knowledge sharing in emergency events: a virtual community perspective. Syst. Res. Behav. Sci. **30**(3), 367–382 (2013)

14. Jin, J., et al.: Why users contribute knowledge to online communities: an empirical study of an online social Q&A community. Inf. Manag. **52**(7), 840–849 (2015)

15. Fu-ren, L., Hui-yi, H.: Why people share knowledge in virtual communities?: The use of Yahoo! Kimo Knowledge as an example. Internet Res. Electron. Networking Appl. Policy **23**(2), 133–159 (2013)

16. Jin, X.-L., et al.: Why users keep answering questions in online question answering communities: a theoretical and empirical investigation. Int. J. Inf. Manag. **33**(1), 93–104 (2013)

17. Martínez-Torres, M.R., et al.: A text categorisation tool for open source communities based on semantic analysis. Behav. Inf. Technol. **32**(6), 532–544 (2013)

18. Liao, X., Li, Z., Xi, Y.: Modeling and analyzing methods of user-innovation knowledge in enterprise communities based on weighted knowledge network. Syst. Eng. Theor. Pract. **2016**(1), 94–105 (2016)

19. Watts, D.J., Strogatz, S.H.: Collective dynamics of 'small-world' networks. Nature **393** (6684), 440–442 (1998)

20. Barabási, A.L., Albert, R.: Emergence of scaling in random networks. Science **286**(5439), 509–512 (1999)

21. Zhou, H.-R., Ma, Y.-P., Ma, Y.-Z., et al.: Overview of development of network science. Comput. Eng. Appl. **45**(24), 7–10 (2009)

22. Newman, M.E.J.: Scientific collaboration networks. I. Network construction and fundamental results. Phys. Rev. E **64**(1), 016131 (2001)

23. Newman, M.E.J.: Scientific collaboration networks. II. Shortest paths, weighted networks, and centrality. Phys. Rev. E **64**(1), 016132 (2001)

24. Klemm, K., Eguiluz, V.M.: Highly clustered scale-free networks. Phys. Rev. E **65**(3), 036123 (2002)

25. West, G.B., Brown, J.H., Enquist, B.J.: A general model for the origin of allometric scaling laws in biology. Science **276**(5309), 122–126 (1997)

26. Watts, D.J.: A simple model of global cascades on random networks. Proc. Natl. Acad. Sci. **99**(9), 5766–5771 (2002)

27. Cohen, J., Briand, F., Newman, C.: Community Food Webs: Data and Theory. Springer, Heidelberg (2012)

28. Seufert, A., Von Krogh, G., Bach, A.: Towards knowledge networking. J. Knowl. Manag. **3** (3), 180–190 (1999)

29. Xi, Y.-J., Dang, Y.Z.: The discovery and representation methods of expert domain knowledge based on knowledge network. Syst. Eng. **23**(8), 110–115 (2005)

30. Xi, Y.-J., Dang, Y.Z.: Knowledge supernetwork model and its application in organizational knowledge systems. J. Manag. Sci. China **12**(3), 12–21 (2009)

31. Xi, Y.-J., Dang, Y.Z.: Method to analyze robustness of knowledge network based on weighted supernetwork model and its application. Syst. Eng. Theor. Pract. 27(4), 134–140 (2007)
32. Colazo, J.: Performance implications of stage-wise lead user participation in software development problem solving. Decis. Support Syst. 67(C), 100–108 (2014)
33. Schweisfurth, T.G., Raasch, C.: Embedded lead users—The benefits of employing users for corporate innovation. Res. Policy 44(1), 168–180 (2014)
34. Franke, N., Hippel, E.A.V., Schreier, M.: Finding commercially attractive user innovation: a test of lead-user theory. J. Prod. Innov. Manag. 23(4), 301–315 (2005)
35. Marchi, G., Giachetti, C., Gennaro, P.D.: Extending lead-user theory to online brand communities: the case of the community Ducati. Technovation 31(8), 350–361 (2011)
36. Füller, J., Matzler, K., Hoppe, M.: Brand community members as a source of innovation. J. Prod. Innov. Manag. 25(6), 608–619 (2008)
37. Jeppesen, L.B., Laursen, K.: The role of lead users in knowledge sharing. Res. Policy 38 (10), 1582–1589 (2009)
38. Hung, C.L., Chou, J.C.L., Shu, K.Y.: Searching for lead users in the context of web 2.0. In: IEEE International Conference on Management of Innovation and Technology, pp. 344–349 (2008)
39. Hippel, E V.: Innovation by User Communities: Learning From Open-Source Software. Mit Sloan Manag. Rev. 42(4), 82–86 (2001)
40. Franke, N., Shah, S.: How communities support innovative activities: an exploration of assistance and sharing among end-users. Res. Policy 32(1), 157–178 (2003)

A Conceptual Model of Optimizing Service System for the 3rd Generation Service Innovation

Michitaka Kosaka[1(✉)] and Jing Wang[2]

[1] Japan Advanced Institute of Science and Technology, Nomi, Japan
kosa@jaist.ac.jp
[2] School of Economics and Management, Beihang University, Beijing, China
jim08@buaa.edu.cn

Abstract. Information and communication technology (ICT) has been influencing on service innovation. The 1st generation service innovation has been developed using technologies such as customer database in intranet business information systems. The 2nd generation service innovation has been brought by the internet. Now we are in the new information technology era using Internet of things (IoT) or artificial intelligence (AI) such as deep learning. This paper proposes a concept of the 3rd generation service innovation, which optimizes service system creating service values for customers.

Keywords: Service innovation · Value creation · Service system · Optimization · Internet of things

1 Introduction

Information and communication technology (ICT) has been contributing to various service innovation, where ICT has been changing services and has an important role for customers' value creation and providers' profit generation. So far, ICT revolution can be categorized to three generation. The 1st generation is an intranet era. Companies introduced enterprise information systems, streamlined their work and collected customers' information for creating customers' values. Point-of-sales (POS) systems in distribution industries, online banking systems in financial industries are typical examples. The 2nd generation is an internet era. The internet connects service providers and customers all over the world for 24 h and 365 days a year. The internet has brought various kinds of services such as travel service mediators or information retrieval services. Now, the third generation ICT revolution has been discussed, where new technologies such as Internet-of-things (IoT), Deep learning are employed. These new technologies will change services toward the 3rd generation service innovation. The 3rd generation service innovation is expected to provide solutions to various issues in the 21st century such as manufacturing servitization, care for aging society, and so on. Service science for the 3rd generation service innovation should pursue service value creation using advanced technologies in ICT revolution, which are Internet of things (IoT), Deep learning, cloud computing environment, and so on.

© Springer Nature Singapore Pte Ltd. 2016
J. Chen et al. (Eds.): KSS 2016, CCIS 660, pp. 68–81, 2016.
DOI: 10.1007/978-981-10-2857-1_6

In this paper, we propose a new direction of service science for the 3rd generation service innovation. First, we review the history of service innovation and service value creation, and show that ICT gives the significant impact on service value creation in each service innovation. Next, we consider the service value creation using advanced technologies in the 3rd generation service innovation. There is the most important question in this new direction, that is, what is the essential mechanism of value creation using IoT or deep learning? In order to give the answer to this question, it is shown that the structure of service value creation can be described as a service system and the optimization of this service system is essential for service value creation in the 3rd generation service innovation. This is the direction of not only the 3rd generation service science but also the 3rd generation system science.

2 Three Generations of Service Innovation Dependent on the Progress of ICT

Various service sciences and technologies have been researched for service innovation. They are service marketing [1], service management, IT service management, Web service. Recently, service system research [2] is remarked. These researches are deeply related to ICT progress. From the viewpoint of ICT progress, we can categorize service researches and service innovation into three generations as shown in Fig. 1.

The 1st generation service innovation has been developed by using intranet information systems with network and customers' database and aimed at enhancement of service quality in traditional services such as hotel services, transportation services, maintenance services and so on. The service marketing theory is used as a theoretical framework for service value creation.

The 2nd generation service innovation has been developed under the internet environment, where services can be utilized for 24 h and 365 days a year all over the

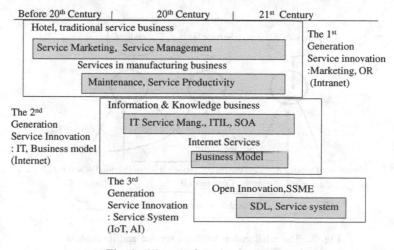

Fig. 1. History of service innovation

world. The service value creation is done in the cyberspace and new business models such as long-tail business model have been developed. Also, the service concept is applied to develop new information technologies such as service oriented architecture (SOA) or Web services.

The 3rd generation service innovation has just started according to new ICT such as IoT or AI (deep learning). At the same time, new service research such as service dominant logic (SD-logic) [3] claims the importance of value co-creation or resource integration for creating service value. Considering these trends, the research for the 3rd generation service innovation should clarify the role of new ICT for service value creation.

3   Service Value Creation Using ICT in Each Generation

3.1   Basic Mechanism of Service Value Creation

Service value depends on the relationship between the provided service and its context which shows the necessity of the service. Even if the quality of the service is high, the service value is determined based on users' needs. Service value is dependent on customers' contexts corresponding to the provided service. This is a similar concept as "Value-in-use" in SD-logic. This relationship can be shown in Fig. 2 [4].

In Fig. 2, Service A has high value for Customer A who wants it but does not have value for Customer B or Customer C who does not want it. Service innovation pursuits to create or enhance service value based on this characteristic of service value creation. That is to say, the context of service should be investigated and the relationship between service and its context should be clarified. What was the role of ICT in the first and the second generation service innovation? What is the role of ICT in the third generation service innovation? How can we create service value using ICT? These are important research questions about the role of ICT in service innovation.

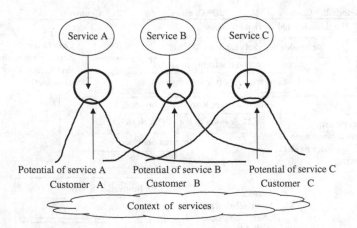

Fig. 2. The relationship between service and its context

3.2 Service Value Creation in the 1st Generation Service Innovation

(1) Examples of the 1st generation service innovation

The 1st generation service innovation has been realized based on the research of service marketing and the utilization of intranet system. The service marketing mainly consists of operational research (OR) and management. It generated lots of methodologies in the 20th century. Also, most of all companies introduced their enterprise intranet systems for improving business performance. They utilize online networks and customer/sales database. There are many examples of the 1st generation service innovation such as POS systems in distribution industries and yield management systems in airline companies or hotels, which are explained as follows.

a. POS system in Seven-eleven convenience store

POS system of Seven eleven is a successful example of improving service value by using ICT. In this system, customers' needs at any timing in each store are predicted by analyzing POS sales data and suitable products satisfying their needs are provided from the distribution center to each store. This means that ICT reveals customers' context related to their purchasing products. This system can improve service productivity by providing small amounts and large kinds of goods.

b. Yield management system in airline companies

In airline companies, yield management systems have been introduced in order to optimize their profit. The system provides the optimal price planning by changing prices according to the low season and the high season. This can be done by combining mathematical optimization methodologies in OR and customer sales database.

(2) The role of ICT in the 1st generation service innovation – deep understanding the customer's context of the provided services

In the 1st generation service innovation, service value is generated by identifying the customers' context using customers' sales data collected by intranet systems. The service value creation process is shown in Fig. 3. First, the context is identified by using various methodologies and data. Then the suitable service corresponding to the context is provided. Various data such as interview data, questionnaire data, sales data are utilized for identifying customers' context. ICT such as data mining is more powerful methodology because ICT can utilize large volume of data and provide proper context related to customers' requirements in real time.

Thus, the role of ICT in the 1st generation service innovation is to understand the customers' context related to the provided services. If we understand the right relationship between service and its context, we can provide high service value to customers as shown in Fig. 2. Many examples such as Seven-eleven POS system show the effectiveness of ICT to improve their service productivity and their profit.

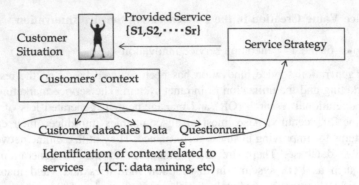

Fig. 3. Service value creation process

3.3 The 2nd Generation Service Innovation

(1) Examples of the 2nd generation service innovation

The 2nd generation service innovation was developed due to the internet, which brought the business environment of 24 h and 365 days a year. Services have been changed dramatically and service value has been created at anytime and anywhere in the cyber space. Therefore, new service business models using the internet have been developed. The long tail business model and the service mediator business model are representative examples.

a. Long tail business model

Services in the internet environment can be accessed from all over the world. This means that customers and service providers are considered to be unlimited. The business focusing on the specific customer target can survive by collecting customers all over the world. This business model is called as the long tail business model as shown in Fig. 4. The success factor is connection between specific services and customers who want them in the internet environment.

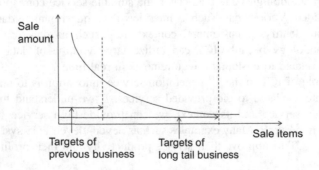

Fig. 4. Long tail business model

b. Internet service mediator

There are various customers and service providers in the internet environment. They have no knowledge about what services are provided and what customers exist. Therefore, new service value is generated by connecting customers and suitable services. This is the connecting service value by the internet service mediators as shown in Fig. 5. The representative example is Airbnb which connects lodgings and travelers who have same value feelings all over the world.

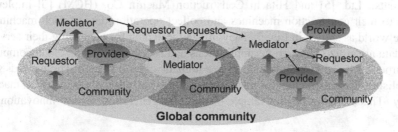

Fig. 5. Service mediator in global community

(2) The role of ICT in the 2nd generation service innovation – service values by connecting new services and new customers

In the second generation service innovation, the essential value is the connecting value of new services and new customers as shown in Fig. 6. Based on this new value creation, various new service business models such as long tail business model, service mediator business model have been developed in the internet environment. As shown in Fig. 6, the relationship between new services and new customers' contexts can be revealed by removing the restriction of time and space in the internet environment. This is the mechanism of service value creation in the 2nd generation service innovation.

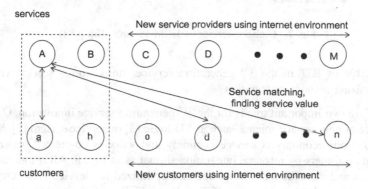

Fig. 6. Service value by connecting in the Internet environment

3.4   The 3rd Generation Service Innovation

(1) Examples of the 3rd generation service innovation

The third generation service innovation has just started. Therefore we have little examples. However, Machine to Machine (M2M) technologies have been already used in construction machine industry.

a. Utilization of IoT in construction machine industry

Komatsu Ltd. [5] and Hitachi Construction Machin Co. (HCM) [6] implement sensors to their construction machines and collect operation data of each machine all over the world as shown in Fig. 7. They utilize these data for enhancing their services. These data are utilized for various applications and services under the cloud computing environment. Also, they are looking at the automatic control system. These trends seem to establish ICT infrastructure for manufacturing servitaization of construction machine industry. This seems to be an example of the third generation service innovation.

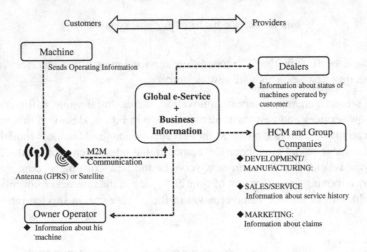

Fig. 7. Global e-service of HCM (Source: [6], p.83)

(2) The role of ICT in the 3rd generation service innovation – service values by optimizing resource integration

There are two important trends for the 3rd generation service innovation. One is the appearance of new service science such as SD-logic [3] or service system [2]. SD-logic claimed that our economy is service economy and value is co-created between providers and customers by resource integration shown as in Fig. 8. All resources related to providers and customers are integrated and optimized for service value creation.

Another important trend is a new ICT such as IoT, Deep learning, cloud computing. All things are connected in the Internet environment and various data and AI technologies such as deep learning are utilized for service value creation through resource

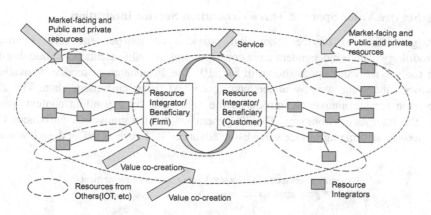

Fig. 8. Resource integration in service dominant logic

integration. These two important trends are effective for providing solutions to recent issues such as manufacturing servitization or healthy aging community, where optimization of service system is expected using advanced technologies. This is the 3rd generation service innovation and service value is obtained by optimizing service system using advanced ICT.

Especially, IoT seems to be an important factor revolutionizing service business as well as the Internet changed service business as the 2nd generation service innovation. There are several sprouts of the 3rd generation service innovation using new technologies. By using IoT, we can understand changing users' context in real time. Also, AI such as deep learning can optimize service according to changing user's context.

Value co-creation between providers and customers can be done through resource integration using IoT and AI. The most important thing is to create high service value through optimization of service system. This characteristic can be shown in Fig. 9. From the viewpoint of service value, the relationship between provided service and its context is kept to be optimum by using advanced technologies such as IoT, AI, robots and so on.

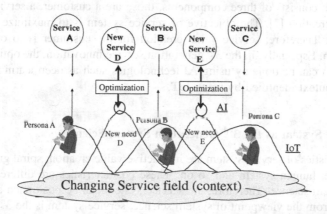

Fig. 9. Optimization of the relationship between context and provided service

3.5   Service Value Space of Three Generation Service Innovation

Each generation of service innovation provides an independent value creation methodology. Three independent axes corresponding to value creation methodologies form the service value space shown in Fig. 10. The 1st generation service innovation introduced the value axis by understanding context using customers' data. The 2nd generation service innovation introduced the value axis by expanding context space using the Internet environment. The 3rd generation service innovation introduces the value axis by optimizing service value using advanced ICT such as IoT or deep learning.

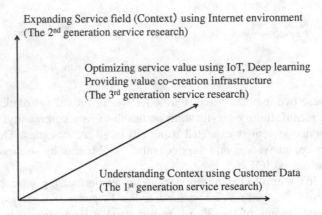

Expanding Service field (Context) using Internet environment
(The 2nd generation service research)

Optimizing service value using IoT, Deep learning
Providing value co-creation infrastructure
(The 3rd generation service research)

Understanding Context using Customer Data
(The 1st generation service research)

Fig. 10. Value space created by using ICT

4   Service Value Creation as Optimization of Service System from the Viewpoint of System Science

4.1   Service Value Creation as the Optimization of Service System

According to our definition of service, service can be described as a service system as Fig. 11, which consists of three components, those are a customer, a service provider and their co-creation [7]. The objective of service system is to maximize customer's service value. Therefore, service value creation for the customer is to optimize the service system. Especially, in the 3rd generation service innovation, the optimization of service system can be done by using AI technologies such as deep learning based on customers' context identified by using IoT.

4.2   Service System as the 3rd Generation System Science

The characteristics of service system are interactive value creation, spiral growth based on experience, human system and so on. These characteristics are different from the previous system science. Therefore, the service system can be positioned as shown in Fig. 12 [8]. From the viewpoint of system science, service system is the 3rd generation system science after hard system science (1st generation) and soft system science

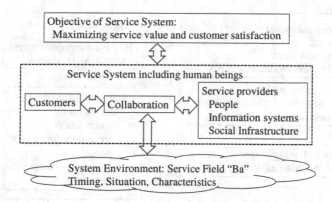

Fig. 11. Optimization of Service system including human beings

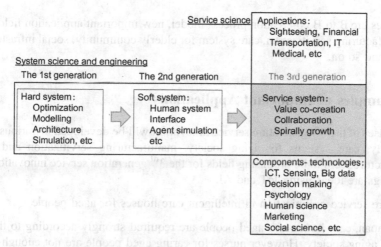

Fig. 12. Service System as the third generation system science

(2nd generation). From the viewpoint of service science, service system is the core for service value creation using technologies in various services. The research objective of service system is how to organize service system architecture and control value co-creation process. If we establish the service system architecture and the process of service value creation in the service system, we can utilize various advanced ICT for realizing it and apply the service system to various application fields.

4.3   Architecture Model for the 3rd Generation Service Innovation

Based on the 3rd generation system science, the theoretical model and application research for the 3rd generation service innovation can be summarized as shown in Fig. 13. The theoretical model consists of three layers, which are service system architecture layer, new technology layer, and business layer. Important application

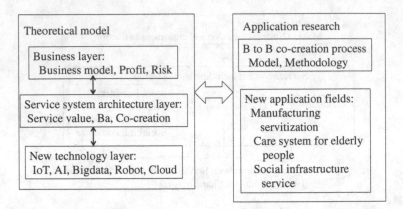

Fig. 13. Architecture for the 3rd generation service innovation

researches are B to B co-creation process model, new important application fields such as manufacturing servitization, care system for elderly community, social infrastructure service and so on.

5 Examples of Important Applications

Many cases of the 3rd generation service innovation will be developed in various fields. Especially, care systems for aging society, manufacturing servitization and social infrastructure seem to be promising fields for the 3rd generation service innovation. The followings are ideas using IoT and AI.

(1) Care service optimization in intelligent care houses for aged people

In Japan, care houses for aged people are required strongly according to the progress of aging society. However nurses for caring aged people are not enough due to decrease in the number of young workers. Therefore it is necessary for care houses to reduce nurses' load and improve service quality for aged people by using advanced technologies such as IoT and AI.

Generally, care plans for services in 1 week are prepared based on the situation or needs of aged people. Care services are provided based on this 1 week care plan. The care plan is very important for service quality and special knowledge based on nurse experiences is need for making a good care plan. However, nurses do not have enough time for making good care plans because they are busy for caring aged people. The intelligent care plan system using advanced technologies is required for reducing nurses' works and improving care service quality.

Figure 14 is an idea of an intelligent care plan. Various sensors are used for observing aged people situation and send collected data to the care plan system. Also, aged people provide their needs for care services, their levels of required care, and their daily life pattern. By using such information, the care plan system can identify the contexts of care services for aged people. On the other hand, care services include various services such as meal service, bath services and so on. The care plan system can optimize automatically 1 week care services by using AI based on the contexts identified using collected data. The combinatorial optimization of candidate services is done in order to maximize the service values of care service receivers under the current situation, This is an idea of combining IoT and AI for intelligent care houses. This system reduces nurses' work load and improves quality of care services.

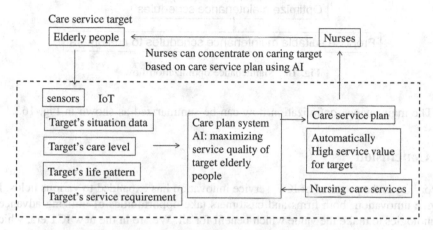

Fig. 14. Care plan optimization in care house for aged people

(2) Optimization of maintenance in construction machine industry

The maintenance management in construction machine industry has been studied for effective maintenance using IoT system as shown in Fig. 7. Figure 15 compares the traditional maintenance process and a new maintenance process optimized using IoT. If IoT can observe situation of all machines and parts in the maintenance process, then the most effective maintenance schedules for customers and providers might be calculated based on predicted failures of machines.

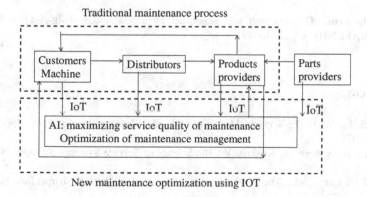

Fig. 15. Value chain management in manufacturing servitization

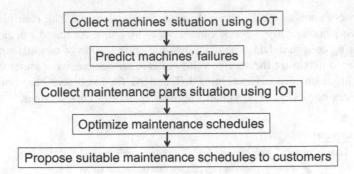

Fig. 16. Maintenance optimization flow

The maintenance optimization flow can be summarized as shown in Fig. 16.

6 Conclusion

According to the progress of ICT, service innovation has expanded to various fields. In service innovation, both firms and customers take opportunities of utilizing advanced technologies to make the value which benefit for all. We are in the new ICT era, which is beyond the internet era by using IoT and deep learning. In this IoT era, a new service innovation using advanced technologies is expected.

This paper proposes a new concept of the 3rd generation service innovation that optimizes the service system for creating the service value. The concept provides a mechanism for firms and customers to co-create service value effectively. Moreover, we propose that the 3rd generation system science is the service value co-creation system and the optimization of service system is important research issues there.

In fact, this research is a conceptual proposal. There are many hopeful applications for the 3rd generation service innovation, which are care systems for aging society, manufacturing servitization and social infrastructure. Future researches are expected to demonstrate the effectiveness of the proposed concept.

Acknowledgement. This research was supported by Japanese Grants-in-Aid for Scientific Research (KAKENHI No. 25240049).

References

1. Lovelock, L., Wirtz, J.: Services Marketing, 6th edn. Pearson Prentice Hall, Upper Saddle River (2007)
2. Demirkan, H., Spohrer, J., Krishna, V.: The Science of Service Systems. Springer, New York (2011)
3. Lusch, R.F., Vargo, S.L.: The Service Dominant Logic of Marketing. Sharpe, Inc., ME (2006)

4. Kosaka, M., Shirahada, K., Ito, Y.: A concept of service field in service systems for creating service value. In: Proceedings of the 4th Japan-China Joint Symposium on Information Systems, Nanchang, Jiangxi, China, pp. 35–40 (2011)
5. Sumi, T., Kitatani, T.: Trends and issues in service business innovations in Japanese manufacturing industry. In: Progressive Trends of Knowledge and System Based Science for Service Innovation, pp. 237–257. IGI global (2013)
6. Matsuda, F., Kosaka, M.: Hitachi Construction Machine Co., Ltd.- M2M and cloud computing based information service. In: Wang, J., Kosaka, M., Xing, K. (eds.) Manufacturing Servitization in the Asia-Pacific, pp. 75–92. Springer, Singapore (2016)
7. Kosaka, M.: A service value creation model and the role of ethnography. In: An Ethnography of Global Landscapes and Corridors, pp. 109–130. InTech (2012)
8. Kosaka, M.: A framework for service system research based on system's approach. In: The 4th Asian Conference on Information Systems (ACIS 2015), Penang, Malaysia (2015)

A Methodology for Problem-Driven Knowledge Acquisition and Its Application

Yin Gai[1(✉)], Yanzhong Dang[2(✉)], and Zhaoguang Xu[2]

[1] School of Management Science and Engineering,
Dongbei University of Finance and Economic,
Dalian, Liaoning Province, People's Republic of China
gaiyin@dufe.edu.cn
[2] Institute of System Engineering, Dalian University of Technology,
Dalian, Liaoning Province, People's Republic of China
dangyanzhong@dlut.edu.cn

Abstract. Most papers claim that knowledge acquisition is a critical bottleneck and has become an important research issue in knowledge management. However they take very different perspectives on it and there is little agreement on how it should be implemented in the industry. In this paper, we investigate problem solving as a cognitive process and propose a new and practical methodology for problem-driven knowledge acquisition. Our contribution includes: (1) proposing a model for problem-driven knowledge acquisition at cognitive level which integrates cognitive structures of problem solvers with situations of knowledge acquisition; (2) designing a framework for problem-driven knowledge acquisition which implements knowledge structured from prior practical experience; (3) developing a prototype system and implementing a mechanism of problem-driven knowledge acquisition. Our proposed methodology has been applied in a real practice case. The results demonstrate that the prototype system can be an effective tool for enterprise-wide knowledge management practice.

Keywords: Problem-driven · Problem solving · Knowledge acquisition · Knowledge management

1 Introduction

Knowledge undoubtedly represents the main competitive advantage of an organization. It refers to what one knows and understands. It is "meaningful links people make in their minds between information and its application in action in a specific setting" [1]. That is to say, knowledge can be viewed both as a thing to be stored and manipulated and as a process of simultaneously knowing and acting [2]. In the industry, knowledge refers to the sum of information relevant to a certain job, and usually we get things done successfully by knowing an answer or how to find an answer, or knowing someone who can [3]. Since knowledge has become important for individuals and organizations increasingly, knowledge management has been widely proposed as a methodology that can manage knowledge in organizations [4]. The idea about knowledge management is simple: apply knowledge to a work environment in order to create value.

© Springer Nature Singapore Pte Ltd. 2016
J. Chen et al. (Eds.): KSS 2016, CCIS 660, pp. 82–93, 2016.
DOI: 10.1007/978-981-10-2857-1_7

In this context, acquiring and creating knowledge is a key issue to improve the performance of organizations over time. However, acquiring knowledge is difficult in organizations for three main reasons: (1) knowledge is often implicit – it lives in the minds of individuals. Therefore, it is difficult to transfer knowledge to another person by means of the written word or verbal expression; (2) knowledge is situated – it is created in a context, so it cannot be used reliably out of context where it is created; (3) knowledge acquisition is a cognitive process – it involves both dynamic modeling and knowledge generation activities.

In order to overcome these difficulties, a wide variety of organizational knowledge management practices have been proposed to support knowledge acquisition, storage and creation involving ideas from different research areas, such as psychology, sociology, philosophy, and computer science [5, 6], yet the process how humans acquire knowledge is not clear [7, 8].

All life is problem solving [9]. Problem solving is in the center of daily operations for organizations. Workers are required to organize complex projects, deal with interpersonal conflicts, and develop innovative products. Since 1980s, problem solving is considered as an important, if not the most essential, feature of learning by many instructional models like Constructivist Learning Environments, or Problem-based Learning [10]. In the decades, problem solving that requires searching and sharing knowledge among a group of actors in a particular context becomes an important issue in knowledge management [11, 12]. Problem solving may enable or inhibit an organization's or an individual's ability on problem solving [13–15]. The researches of problem solving may provide some guidance for the design of organizations to support knowledge acquisition and generation [16–18].

By focusing attention on the importance of problem solving as a cognitive process, this study proposes a new way to understand the connection between problem solving and knowledge acquisition as well as a methodology for problem-driven knowledge acquisition. The methodology is connected with: (1) accumulating and formalizing prior knowledge acquisition at cognitive level, (2) presenting a cognitive knowledge model for a new problem solving, and (3) integrating new knowledge into existing knowledge and experiences. The main goal of this study is in two folds. First, it aims to introduce a framework that can be used for knowledge acquisition driven by problem solving in daily work. Second, it aims to develop a problem-driven knowledge management methodology to support organizational knowledge management practice. Our work differs from existing literature in the following ways: (1) it incorporates cognitive structures of problem solvers with situations of problem solving into the knowledge acquisition model, (2) it analyzes how this methodology can be successfully applied into enterprise-wide knowledge management practice.

This paper is organized as follows. Section 2 analyzes the process of problem solving in coherence with the theories of cognitive psychology and knowledge management practice. Section 3 proposes a model for problem-driven knowledge acquisition at cognitive level, and designs a framework consists of two phases with six steps correspondingly. An application case of the methodology for problem-driven knowledge acquisition is provided in Sect. 4. Finally, Sect. 5 concludes and presents some future research extensions along with this work.

2 Problem Solving

From the view of cognitive psychology, there are many theories proposed about the cognitive process of problem solving [19, 20]. Generally, the process of problem solving is divided into several stages with different features in each stage. Based on these theories about the process of problem solving and knowledge practice, in this paper the process of problem solving is divided into five stages: Describing, Analyzing, Proposing, Implementing and Evaluating, as shown in Fig. 1. In fact, the process of problem solving is often in a spiral pattern so that the problem can be solved completely.

Fig. 1. The five stages of problem solving process

Stage (1): Describing

Problem solving usually begins with some description about the problem situations, as well as the initial state and the terminal state. The problem situations include the time, space and characters during the process of problem solving; initial state includes the conditions or phenomena that can be directly observed or tested when the problem happened; the terminal state includes the expected states that can be achieved when the problem is solved completely.

Stage (2): Analyzing

Causes associated with the problem are analyzed and the successive causality among them is then to be traced. Usually, the causes associated with the problem should be reviewed and analyzed from the surface cause to the root one. In other words, analyzing these causes is a process from the outside cause to the inside one, and from the shallower cause to the deeper one about problem solving.

Stage (3): Proposing

Proposing the schema is to solve a specific problem and to give the specific plans. Usually, the proposed schema is to find a variety of solutions according to the causes after the analyzing stage (2). In general, appropriate solutions are proposed for the root causes correspondingly in order to solve the problem completely. And for a relatively complex problem, the root causes may not be single.

Stage (4): Implementing

In fact, many problems could not be simply implemented according to the schema from stage (3), because each problem solver has an individual's preferences and traits. Besides, it must be in line with the incorporation of information from multiple sources of problem solvers' knowledge, perspectives, and experiences, as well as the enterprise's resource that can be obtained.

Stage (5): Evaluating

Although, the problems get solved from stage (1) to stage (4) or expected states are finally achieved, the knowledge or experience during the process of problem solving should be evaluated that may be applied into the subsequently problem solving. So that, the knowledge or experience will continue to promote by the evaluating stage (5).

3   The Methodology

3.1   The Model

In cognitive psychology, problem solving is the cognitive process that implies the efficient interaction between cognitive structures of problem solvers and situations during problem solving. Through analyzing and understanding the cognitive contents of problem solving, two-levels of knowledge are contained at least during the process of problem solving. One is surface-level knowledge which is related to a specific problem, e.g. the details about the problem, the time and space, some issues involved in the process of problem solving, and so on. The other is deep-level knowledge which is general for any problem, e.g. the basic principle, the constraint or the rule, the essence of the problem, and so on. Although there may be differences among different problems, the cognitive structures that has been inherited in the brains of problem solver are common. Therefore, if the surface-level knowledge is abstracted as many situational dimensions and the deep-level knowledge is formalized as a cognitive network respectively, the conceptual model of problem-driven knowledge acquisition is proposed in Fig. 2.

Specifically, in Fig. 2 one situational dimension represents a certain type of situational elements that may be involved in the process of problem solving, such as persons, time, space, and so on. In each situational dimension there is a certain inherent logic

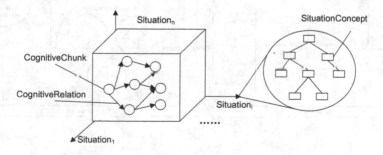

Fig. 2. The conceptual model of problem-driven knowledge acquisition

relation itself. For example, the organizational structure among persons, the sequence of time, the scope and overlapping relations of space, and so on, so that it also can makes more detailed analysis on each situational dimension according to the characteristics of the domain knowledge. The cognitive network is composed of a series of nodes and their relations. One node represents a cognitive chunk, which is a knowledge unit and composed of knowledge points with a certain logical structure. One cognitive chunk has some significant meanings about problem solving. The cognitive chunks are connected through cognitive relations, so that the problem can be transformed from the initial states to the expected states, and finally the problem can be solved.

Generally, the conceptual model integrates the surface-level knowledge which is related to a specific problem with the deep-level knowledge which is general for any problem. The surface-level knowledge is very important for knowledge acquisition, because problem solvers will applied the prior knowledge or experience into any new problem in according to the surface-level knowledge. The deep-level knowledge is inherent in problem solvers' brains which is obtained during the prior process of problem solving.

3.2 The Framework

A framework for problem-driven knowledge acquisition is designed based on the conceptual model proposed in Sect. 3.1. The framework of problem-driven knowledge acquisition consists of two stages with six steps which implement formalization from practical experience to structured knowledge, as shown in Fig. 3.

In the first stage, the practical experience form problem solving is transformed to the semi-structured knowledge. In the second stage, the semi-structured knowledge is formalized to the full-structured knowledge. The following describes the six steps in the framework in detail.

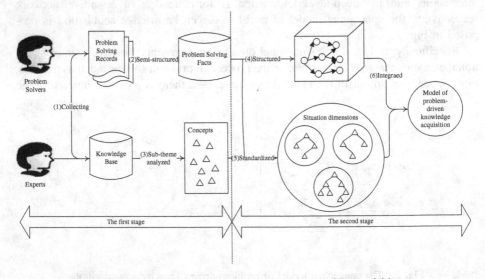

Fig. 3. The framework of problem-driven knowledge acquisition

Step (1) collecting

Under the coordination of decision makers, managers, senior technicians, and the front-line workers, the knowledge engineers endeavor to collect related records during the processes of problem solving and domain knowledge base.

Step (2) semi-structured

According to the five stages about the process of problem solving proposed in Sect. 3, the experience in the records and knowledge base can be semi-structured to problem solving facts.

Step (3) sub-theme analyzed

There are many particular knowledge domains in related to problem solving, so the knowledge base must be analyzed according different aspects about the processes of problem solving. After that, some concepts or words about different knowledge domains are obtained.

Step (4) structured

The problem solving facts are abstracted to a cognitive network with cognitive chunks and their relations. The cognitive network is structured so that the knowledge can be applied into the subsequent problem solving by knowledge retrieval and knowledge organization.

Step (5) standardized

The concepts or words are classified into different situational dimensions, and the standardized terminology is formed. Also, the situational dimensions provide a standardized of the semantics required for knowledge application.

Step (6) integrated

The cognitive network from step (4) is integrated with many situational dimensions from step (5), and the model of problem-driven knowledge acquisition is constructed finally.

4 The Application

This section presents the results of a case study about management practices of the methodology of problem-driven knowledge acquisition, and we have developed a knowledge management prototype system applying the methodology in the workshop for line-stop problems on automatic production lines. We have deliberately simplified the methodology to provide understandable explanations.

4.1 Background

To begin with we will provide a brief background on the case study. This enterprise involved in this practical study is a car manufacturer. In car manufacturing, the stamping process is an important way of metal forming. In order to manufacture mental parts, stamping workshops of modern auto-mobile industry usually use flexible and efficient automatic production lines. There are five automatic production lines in the example stamping workshop, and Line A and Line B are the main production lines. There are various types of stamping molds on Line A and Line B with 2200 stamping molds in the workshop. The productivity of a production line depends not only on how many working strokes, but also on the stop-line time of production line. According to the investigations, the average stop-line rate of production lines in the stamping workshop is as high as 21.6 %, which has brought to the workshop big loss in economy and credibility. Therefore, how to solve stop-line problems on the production lines effectively can not only help the enterprise to reduce production costs, but also meet the urgent needs for just-in-time production.

Usually, when stop-line problems happened, the phenomena on the production line can occur in a variety of ways, and these phenomena often occur overlap together, which have brought great difficulty to restore the production line in a short time. During the process of problem solving, stamping experts are usually able to make correct judgments rapidly by combing pieces of cognitive chunks together in an unconscious process. But when they attempt to explain the problem solving process, it takes often an in-complete and non-sequential form that is not suitable for knowledge acquisition and communication. Even if the relevant participants records the experience, but due to the lack of a structural framework for guidance, important knowledge or experience is difficult for communicating and sharing. And with the expansion of production and update of technology, more and more stop-line problems will occur frequently. Therefore, the aim is to identify technical support for collecting and reusing knowledge and experience about problem solving.

In this project, we employ the methodology for developing a problem-driven knowledge management prototype system so that the workers and managers in the stamping workshop can apply it for providing feedback and guidance to their problem solving and knowledge management.

4.2   The Model of Stop-Line Problem-Driven Knowledge Acquisition

Technicians and experts have much experience about stop-line problem solving from practice over the years. Capturing the experience and knowledge, storing them and distributing them within and across the stamping workshop, are important issues in workshop management. To obtain first-hand information and problem solving experience, we collected the loggings of stop-line problem solving processes from 2012 to 2013, and all the information collected for the project is stored on a local server. Then, we discussed and communicated with experienced technician and experts to understand the stamping process. Finally, we filed 1459 semi-structured records of stop-line problem solving processes stored as Excel sheets, as in Fig. 4. Some English notes have been added in the table headers to describe the problem sheet. Each experience record includes five contents: describing, analyzing, proposing, implementing and evaluating. According to the contents and structures provided in the check sheet, relative participants can express the problem solving process in detail.

Figure 5 shows a cognitive network according to problem solvers' cognitive structures during the process of stop-line problem solving. The cognitive network of the strop-line problem includes two kinds of cognitive chunks: declarative and procedural. Declarative and procedural chunks are separately used to describe some causes and solutions during the process of problem solving. After abstracting from the excel

Problem number 问题编号	When the Problem occurs 问题发现时间	Related departments 来源部门	States of the problem 问题状态	Type1 of the problem 分类1	Type2 of the problem 分类2	Car model 车型	Product variety 生产品种	Production process 工序	Production line 生产线	Production team 班组	Down time 停机时间	Problem title 问题标题
234	2012/2/6	0000	05	02	01	D62A	左侧围		A线	生产乙班（三班）	57	修边切不断
235	2012/2/6	0000	05	02	01	B50	左侧围		B线	生产乙班（三班）	45	op20卡废料频繁
236	2012/2/6	0000	05	02	01	B50	左侧围		B线	生产乙班（三班）	45	op20卡废料频繁
237	2012/2/6	0000	05	02	01	B50	左侧围		B线	生产乙班（三班）	45	op20卡废料频繁
238	2012/2/6	0000	05	02	01	B50	左侧围		B线	生产乙班（三班）	45	op20卡废料频繁
239	2012/2/7	0000	05	02	01	B50	左右支柱内板下部	02	B线	丙段	5	Op20堵废料

Fig. 4. Stop-line problem solving processes stored as excel sheets

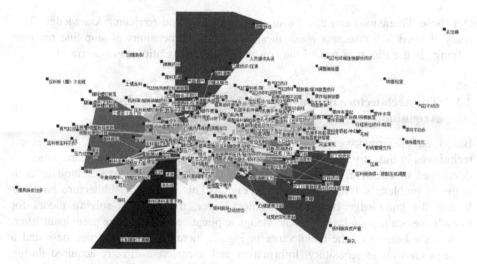

Fig. 5. The cognitive network of stop-line problem solving

sheets, our study get 117 declarative cognitive chunks and 42 procedural cognitive chunks. While circle nodes demotes the declarative cognitive chunks, square nodes denote the procedural cognitive chunks. The thickness of the link between nodes means the probability of the various chunks. The cognitive network offers the visualization for problem analysis and solution.

Figure 6 presents the generalized situations, which encompasses person, equipment, material, technique, and environment situation. Five situational dimensions are constructed for stop-line problem solving on the production lines. A domain expert or manager's role comprises setting up a hierarchy, connecting situational concepts to the dimension. Each situational dimension also can be refined and expanded gradually so

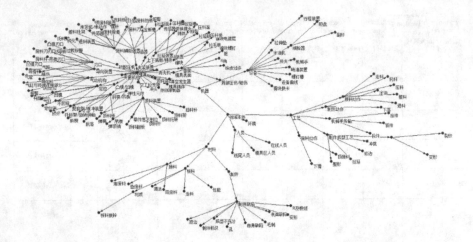

Fig. 6. The situational dimensions of stop-line problem solving (part)

that these dimensions can also be used for indexing and retrieving knowledge. This study defined 519 concepts about these situational dimensions of stop-line problem solving. In the Fig. 6, a part of the concepts and their relationship is showed.

4.3   The Architecture of Stop-Line Problem-Driven Knowledge Acquisition

Based on the cognitive process of problem solving and knowledge management techniques in industry, the architecture for problem-driven knowledge management is developed to effectively support knowledge capitalization and exploitation in each stage of problem solving, as shown in Fig. 7. This designed architecture has four layers: the knowledge base and repository layer, the problem solving model for knowledge acquisition layer, the knowledge application layer and the participant layer.

On the bottom of the architecture in Fig. 7, there are an experience base and a domain knowledge repository. Information and knowledge directly acquired during problem solving are stored to a Problem Solving Experience Base by problem solvers. Each experience is saved as a detailed and semi-structured description of problem solving according to the five stages of problem solving process. Besides, the resources for all the domain information related to problem solving come primarily from the domain knowledge repository, e.g. engineering hand-books, product specification, product functionality, and manufacturing process specification files.

The model for stop-line problem-driven knowledge acquisition is a cognitive network with multi-situational dimensions which incorporates cognitive structures of problem solvers and situations of knowledge acquisition in order to support knowledge acquisition.

Knowledge application layer consists of a problem-solving module, a decision-making module, a knowledge learning module and a knowledge retrieval module. In the problem-solving module, progressive tools are designed to guide problem solvers to

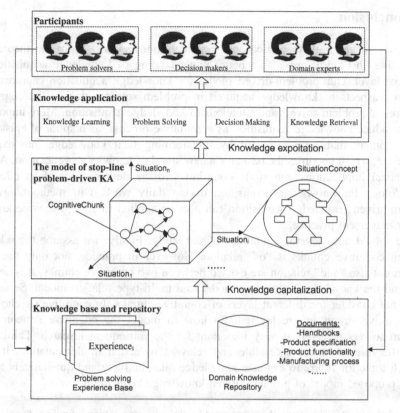

Fig. 7. The architecture of problem-driven knowledge management system

conduct problem solving activities, including requirements description, analysis/ simulation, experiences description, and so on. In decision making module, relative tools help decision makers to collect, analyze, and evaluate knowledge of problem solving. In knowledge learning and retrieval module, the participants can be guided to browse, retrieve, and trace knowledge.

This designed architecture has three types of participants: problem solvers, decision makers and domain experts. Problem solvers represent the users who need obtain some knowledge to solve a problem. They have some basic and general knowledge about the domain. After problem solving, problem solvers need to describe the process of problem solving as an experience record and store to the experience base for knowledge capitalization. Decision makers are the users who make decision. They can rate the process of the problem solving and its components in order to provide feedback for the problem solvers and domain experts. Domain experts are the users who have the deep knowledge and the necessary experience. Their responsibility lies in the assessment of the problem solving process in order to provide a measure of their quality. They can define the concepts of cognitive situations and chunks of cognitive structures through a knowledge editor.

5 Conclusion

This study makes both theoretical and practical contributions. From the theoretical perspective, this study provides a better understanding of knowledge acquisition at cognitive level. Our problem-driven model for knowledge acquisition captures two important aspects of knowledge acquired in problem solving, that is, the cognitive structures of problem solvers and situations of knowledge acquisition. More important is to introduce knowledge acquisition as a cognitive process, as a spiral of epistemological content that grows upward by transforming tacit knowledge into explicit knowledge, which becomes the basis for a new spiral of knowledge generation. As for its practical implications, our study contributes to the development of an effective methodology for knowledge acquisition during daily work. Our methodology for problem-driven knowledge acquisition can also be applied to enterprise-wide knowledge management practice.

We should note several limitations of our study. Firstly, we assume the relation between cognitive chunks is "or" relation. However in practice, not only the "or" relation but also "and" relation are existed between two cognitive chunks. It is needed to extend the knowledge acquisition model to a multi-type relation model. Secondly, we do not consider the different layers of cognitive chunks. In cognitive psychology, the cognitive structures are layered. So how to model the cognitive structures of problem solvers in layers is very important for the subsequent research. Thirdly, to make the research more accessible and relevant to actual implementation, it is a research topic for future to evaluate knowledge quantitatively and qualitatively based on the proposed model of problem-driven knowledge acquisition.

Acknowledgement. This research is supported by the following research funding: the National Natural Science Foundation of China (71031002, 71501032), and the Social Science Foundation of Ministry of Education of China (14YJC630036, 15YJC630193), respectively.

References

1. Edwards, M.J.: Common knowledge: how companies thrive by sharing what they know. Long Range Plan. **34**, 872 (2001)
2. Robins, S.: Mindreading and tacit knowledge. Cogn. Syst. Res. **28**, 1–11 (2014)
3. Lin, Y.C., Lee, H.Y.: Developing project communities of practice-based knowledge management system in construction. Autom. Constr. **22**, 422–432 (2012)
4. Gebus, S., Leiviskä, K.: Knowledge acquisition for decision support systems on an electronic assembly line. Expert Syst. Appl. **36**, 93–101 (2009)
5. Motta, E.: 25 years of knowledge acquisition. Int. J. Hum. Comput. Stud. **71**, 131–134 (2013)
6. Gruber, T.R.: Nature, nurture, and knowledge acquisition. Int. Hum. Comput. Stud. **71**, 191–194 (2013)
7. Brian, R.G.: Knowledge acquisition: past, present and future. Int. J. Hum. Comput. Stud. **71**, 135–156 (2013)

8. Aussenac-Gilles, N., Gandon, F.: From the knowledge acquisition bottleneck to the knowledge acquisition overflow: a brief French history of knowledge acquisition. Int. J. Hum. Comput. Stud. **71**, 157–165 (2013)
9. Popper, K.: All Life Is Problem Solving. Routledge, London (1999)
10. Breuker, J.: A cognitive science perspective on knowledge acquisition. Int. J. Hum. Comput. Stud. **71**, 177–183 (2013)
11. Chen, Y.-J.: Development of a method for ontology-based empirical knowledge representation and reasoning. Decis. Support Syst. **50**, 1–20 (2010)
12. Jaques, P.A., Seffrin, H., Rubi, G., Morais, F.D., Ghilardi, C., Bittencourt, I.I., Isotani, S.: Rule-based expert systems to support step-by-step guidance in algebraic problem solving: the case of the tutor PAT2Math. Expert Syst. Appl. **40**, 5456–5565 (2013)
13. Kubota, F.I., Rosa, L.C.D.: Identification and conception of cleaner production opportunities with the theory of inventive problem solving. J. Cleaner Prod. **47**, 199–210 (2013)
14. Hao, J.X., Kwok, C.W., Lau, Y.K., Yu, A.Y.: Predicting problem-solving performance with concept maps: an information-theoretic approach. Decis. Support Syst. **48**, 613–621 (2010)
15. Yu, W.D., Yang, J.B., Tseng, J.C.R., Liu, S.J., Wu, J.W.: Proactive problem-solver for construction. Autom. Constr. **19**, 808–816 (2010)
16. Jabrouni, H., Foguem, B.K., Geneste, L., Vaysse, C.: Continuous improvement through knowledge-guided analysis in experience feedback. Eng. Appl. Artif. Intell. **24**, 1419–1431 (2011)
17. Aarikka-Stenroos, L., Jaakkola, E.: Value co-creation in knowledge intensive business services: a dyadic perspective on the joint problem solving process. Ind. Mark. Manag. **41**, 15–26 (2012)
18. Mast, J.D., Lokkerbol, J.: An analysis of the Six Sigma DMAIC method from the perspective of problem solving. Int. J. Prod. Econ. **139**, 604–614 (2012)
19. Wang, Y., Chiew, V.: On the cognitive process of human problem solving. Cogn. Syst. Res. **11**, 81–92 (2010)
20. Reimann, P., Kickmeier-Rust, K., Albert, D.: Problem solving learning environments and assessment: a knowledge space theory approach. Comput. Educ. **64**, 183–193 (2013)

A Framework for Analyzing Vulnerability of Critical Infrastructures Under Localized Attacks

KeSheng Yan[(✉)], LiLi Rong, Tao Lu, and ZiJian Ni

Institute of Systems Engineering, Dalian University of Technology,
Dalian 116024, People's Republic of China
yan_kesheng@126.com, {llrong,lutao}@dlut.edu.cn,
nizijian@hotmail.com

Abstract. Critical infrastructures (CIs) are usually spatially embedded, which makes them more susceptible to localized attacks where a set of components located within a localized area are damaged. This paper presents a framework for analyzing vulnerability of CIs under localized attack. Both functional and geographic interdependencies among CIs are considered. Multi-layer networks are employed to establish the functional interdependency. A novel attack failure model, i.e., localized failure model, is proposed to model geographic interdependency and quantify the failure probabilities of components under localized attack. Vulnerability of CIs is investigated from three perspectives, including analyzing the system-level vulnerability, identifying the critical infrastructure components and geographical locations. Our framework can help stakeholders increase their knowledge of vulnerability of CIs under localized attack and protect them more efficiently.

Keywords: Critical infrastructures · Interdependencies · Localized attack · Localized failure model · Cascading failures

1 Introduction

Critical infrastructures (CIs), such as energy supply, transport services, water supply and ICT (information and communication technology) systems, are essential for social well-being and economic development [1]. Currently, CIs are subject to many types of potential threats, such as natural disasters and intentional attacks, which can cause great economic losses and affect a larger number of users [2]. To protect CIs against potential threats, policymakers will benefit from knowledge derived from vulnerability analysis since reducing vulnerability will reduce consequences. Thus, in order to design and protect CIs more efficiently, it requires exploring their response to potential threats and analyzing their vulnerability, which has been a critical field of contemporary research [3–5].

For analyzing vulnerability of CIs, two characteristics need to be considered. First, CIs are often spatially embedded [6], which not only influences their vulnerability dramatically [7] but also makes them even more susceptible to localized

© Springer Nature Singapore Pte Ltd. 2016
J. Chen et al. (Eds.): KSS 2016, CCIS 660, pp. 94–103, 2016.
DOI: 10.1007/978-981-10-2857-1_8

attacks where a set of components distributed in a localized area are damaged [8–11], for example, the effects of earthquake, flood, terrorist or military attacks. Thus, this paper addresses the problem of analyzing the vulnerability of CIs under localized attack. Another characteristic is the fact that, with the development of scientific technology and social economy, CIs have become mutually interdependent. As summarized in Ref. [12], four categories of interdependency exist among CIs: geographic, physical, cyber and logical. In this paper, as analyzed in Sect. 2, physical and cyber interdependencies are jointly referred to as "functional interdependency", and both functional and geographic interdependencies must be taken into consideration in vulnerability analysis of CIs.

In the past decade, many scholars have paid much attention to the issue of vulnerability or robustness analysis of infrastructures [13–18]. For a single infrastructure, probabilistic risk analysis methods and statistical learning theory have been developed to analyze and predict the impact of natural disasters when there is little information about the operation of infrastructure; and network-based approaches have been adopted for modeling infrastructure when its topology is available. For interdependent infrastructures, network-based, agent-based, dynamics-based and economic-based approaches have been proposed to model interdependencies and further analyze the vulnerability [19, 20]. With the development of multi-layer networks tool [21], the network-based approaches have gotten more attention, and many efforts have been done for considering interdependent networks with non-spatial constraint [22] and interdependent spatially embedded networks [7, 23, 24]. However, they all only model functional interdependency among them, ignoring geographic interdependency which must be considered for localized attack. Recently, some literatures have modeled geographic interdependency [8–11]. However, topological neighbor but not geographical distance is used for distance measure in these efforts. Also, only the giant component size is taken as the performance metric to quantify network vulnerability, which have weak correlations to the results obtained when the network flow properties are considered [25]. More importantly, in these studies, components within the affected region caused by localized attack are all failure which do not describe actual damage scenarios. Apparently, a component will fail with certain probability (not always with probability 1).

Despite the recognition of the importance of infrastructure vulnerability and the active researches seen in the past decade, the following seemingly basic questions still remain daunting: (i) how to model both functional and geographic interdependencies among CIs; (ii) how to characterize the failure probability of each component located in the affected region; (iii) how to explore the response of CIs under localized attack. To solve these problems, this paper proposes a framework in which both functional and geographic interdependencies among CIs are modeled: multi-layer network is adopted for modeling the functional interdependency; localized failure model is proposed to model the geographic interdependency and quantify the failure probabilities of components under localized attack. We also verify the feasibility of this framework through analyzing the pertinent vulnerability of artificial interdependent power and water networks.

2   The Vulnerability Analysis Framework

In this section, a framework for the vulnerability analysis of CIs under localized attacks will be introduced, as shown in Fig. 1.

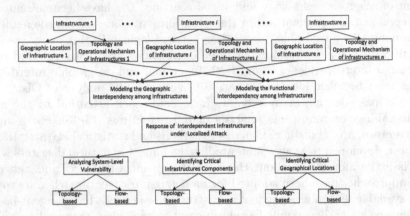

Fig. 1. A framework for the vulnerability analysis of CIs under localized attacks

2.1   Relevant Information Extraction

As shown in the Fig. 1, for analyzing the localized attack vulnerability of CIs, the first step is to extract the relevant information of each infrastructure, including the topology, the geographic location and the operational mechanism.

Most CIs can be represented as network through determining the nodes and links. So, for modeling the CIs of interest, the first issue is to find out what kind of components can be represented as nodes and links. After that, the network topologies of the considered CIs can be naturally obtained. In order to model functional interdependency among CIs, the operational mechanism of each infrastructure is essential, i.e., the information about some components of an infrastructure are interdependent with some components of the other infrastructures. Meanwhile, the geographical information of infrastructures is also required for modeling geographic interdependency among CIs.

Note that, using network-based approach, different authors can model infrastructure from different perspectives, resulting different vulnerability outcomes. In the present framework, following Ref. [26], we choose the generators, the main substations and distribution stations of infrastructure to be modeled as nodes and the edges connected them to be modeled as links, which is the most common modeling method in the related field.

2.2   Modeling the Functional and Geographic Interdependencies

There are mainly four categories of interdependencies among CIs: geographic, physical, cyber and logical [12]. Obviously, the logical interdependency affects slightly the cascading propagation in the short term. Besides, information transmitted can be regarded as "commodity" to the telecommunication infrastructure, which indicates that physical and cyber interdependencies can be jointly referred to as "functional interdependency". For geographic interdependency, this type of interdependency is supposed to be taken into consideration because the components of CIs located in the affected region may damage simultaneously even though they are not functional interdependency [12].

The functional interdependency among CIs can be modeled by inter-links using multi-layer networks, as shown in Fig. 2. Taking power and water systems as example, the functional interdependency between them is that all nodes (generation, transmission and distribution nodes) in water network need electric power provided by some power distribution nodes for their normal operations, and the power generation nodes need water provided by some water distribution nodes for cooling or emissions reduction [12].

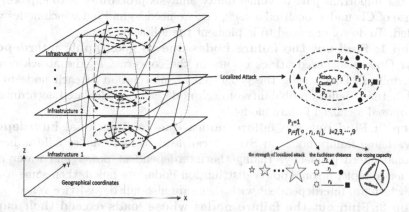

Fig. 2. Localized failure model. Left, CIs are embedded in 3D space. The dashed lines model functional interdependency among CIs, and the solid lines represent the intra-links in each infrastructure. Once a localized attack occurs at a location (for example, a node in Infrastructure 2 is selected as the attack center), it leads to an affected region which is indicated by concentric circles. Right, the nodes in the affected region will fail with a probability which is decided by three indicators.

For geographic interdependency, a new attack failure model, i.e., localized failure model, is proposed in the framework to model it and quantify the failure probabilities of nodes under localized attack. Localized attack is usually confined in a specific geographical region. If the CIs under consideration are located in the affected region, some nodes will be affected and even some of them will fail. Note that one common feature of localized attack is that the attack intensity at a location monotonously declines as its distance from attack center increases. Besides, two nodes perhaps do not fail simultaneously even though they have the same distance from the attack center since they have different coping capacities responding to the attack, which is decided by the node sensitivity, fragility and resilience to localized attack. Thus, as shown in Fig. 2, the failure probabilities of nodes under localized attack can be defined as $P_{\text{failure},i} = f(\sigma, r_i, s_i)$, where σ is the strength of localized attack the CIs suffering from, r_i is the Euclidean distance between node i and the attack center and s_i is the coping capacity of node i. More specifically, considering that Gaussian distribution function mimics as much as possible the downward trend of attack intensity for most types of hazards, we define the function P in Sect. 3 as follows:

$$P_{\text{failure},i} = \left(1 - 2 \int_0^{r_i} \frac{1}{\sqrt{2\pi}} \exp^{\frac{-t^2}{2\sigma^2}} \mathrm{d}t\right)(1 - s_i), \tag{1}$$

where s_i is defined as $s_i = k_i/\max(k_i)$, and k_i is the degree of node i.

2.3 Response of Interdependent CIs Under Localized Attack

The most important part of vulnerability analysis procedures is to explore the response of CIs under localized attack, i.e., we need to find out which nodes will be failed. To do so, one need to implement the following steps.

Step 1: Find out the failure nodes due to geographic interdependency. Once a localized attack occurs in the concern CIs, the attack center is determined. Then, using the geographical information of each node in all infrastructures, failure probability of each node can be calculated according to the proposed localized failure model.

Step 2: Find out the failure nodes due to functional interdependency. Those failure nodes in **Step 1** can lead to more nodes failure due to functional interdependency. Taking the interdependent power and water networks as example, if some power distribution nodes are failed, then some water nodes which are interdependent with them are also failed, and vice versa.

Step 3: Find out the failure nodes whose loads exceed their capacities. Cascading failure can also happen on the overload nodes. The load of failure node will be redistributed to the neighboring nodes, and then some of the neighboring nodes will fail if their actual loads exceed their capacities. Following Refs. [27,28], the initial load of i can be defined as $L_i = k_i^\alpha$, its capacity is $C_i = (1 + \beta)L_i$, and the additional load received by i among the neighboring nodes of failed node j is $\Delta L_{ij} = L_j \frac{C_i}{\sum_{n \in \Gamma_j} C_n}$, where α, β are tunable parameters and Γ_j represents the set of all neighboring nodes of j.

Step 4: Find out the failure nodes which are isolated. In each network, there will be some isolated nodes because their neighboring nodes are all failed. Obviously, this kind of nodes will also be failed.

When a localized attack occurs in the CIs, we can find out all failure nodes by repeating **Step 2-4** until there is no new node failure.

2.4 Vulnerability Analysis from Different Perspectives

The connotation of vulnerability has three related interpretations in the framework [29,30]. In the first interpretation, vulnerability is seen as a system-level property that expresses the extent of adverse effects caused by localized attack. Define system performance in the initial state by P_{initial} and in the post-event state by P_{post}, then vulnerability can be computed as

$$v = \frac{P_{\text{initial}} - P_{\text{post}}}{P_{\text{initial}}}. \tag{2}$$

In order to analyze the system-level vulnerability of CIs, given that it is imprac- tical to set each position of the infrastructure map as localized attack center, as an alternative, we choose to set each node of CIs as localized attack center. Once a localized attack occurs at node i, we can find out the corresponding failure nodes in n infrastructures and calculate the vulnerability of infrastruc- ture l, $v_l, l = 1, 2, \cdots, n$, according to (2). Then its system-level vulnerabil- ity is $V_l = \frac{1}{M} \sum_{l=1}^{M} v_l$, where $M = \sum_{k=1}^{n} N_k$ is the total number of nodes in n infrastructures. In addition, the overall vulnerability of n infrastructures is $V = \sum_{l=1}^{n} w_l V_l$, where $\sum_{l=1}^{n} w_l = 1$.

Different metrics are adopted to characterize and quantify infrastructure per- formance [31], which can be mainly classified into two types: Topology-based that the topology information is only considered and Flow-based that the network flow is further concern. For example, (1) Network Efficiency (Topology-based metric): $NE = \frac{1}{N(N-1)} \sum_{i \neq j} \frac{1}{d_{ij}}$, where N is the size of the network, d_{ij} is the geodesic path distance between i and j; (2) Network Flow Efficiency (Flow- based metric): $FLE = \frac{1}{n_G n_D} \sum_{i \in N_G, j \in N_D} \frac{1}{d_{ij}}$, where N_G and N_D are the sets of generation and distribution nodes respectively, n_G and n_D are their sizes.

In the second interpretation, vulnerability is adopted to describe an infrastruc- ture component [32]. For this interpretation, a component is said to be vulnerable if the failure of it leads to large negative consequences to the infrastructure. The third interpretation of vulnerability is critical geographical locations. Locations where an attack results in large negative consequences would be regard as critical geographical locations [25]. Similarity, the identification of critical infrastructure components and critical geographical locations can be implemented from both Topology-based and Flow-based viewpoints respectively.

To achieve the identifying critical geographical locations, a concept of impact zone is adopted. The whole area the CIs located at is divided into a number of impact zones. An impact zone is a square with a predefined size which is deter- mined based on the area under investigation. We remove the impact zones (all nodes falling in the zone are removed) one by one and calculate the correspond- ing negative consequences. The larger negative consequences a removed impact zone causes, the more critical it is.

3   Simulation Analysis

In this section, in order to demonstrate the feasibility of the proposed vulnera- bility analysis framework, considering that the relevant information of real CIs is confidential, we generate artificial CIs and analyze their vulnerability from three perspectives through simulation.

3.1   Artificial CIs

The proximal topology generator [26] is used for constructing artificial CIs. As shown in Fig. 3, the artificial interdependent power and water networks are generated in area $[0,1] \times [0,1]$. For simulating functional interdependency, we set that the water nodes will be served by the nearest power distribution nodes while the power generation nodes will be driven by the nearest water distribution nodes. In addition, in order to identify the critical geographical locations, the whole area is divided into 25 impact zones.

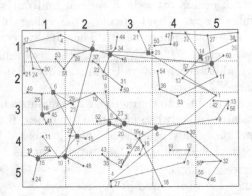

Fig. 3. The power network (the solid red circles, 1–60) and the water network (the solid blue squares, 1–30). The larger circles and squares are the corresponding generation nodes of the two networks, while the leaf nodes are the corresponding distribution nodes. The dash lines divide the whole area into 25 impact zones. (Color figure online)

3.2   Localized Attack Vulnerability Analysis

Firstly, we analyze the system-level vulnerability of the artificial CIs under localized attacks. To do so, we set each node as the localized attack center one by one and find out the corresponding failure nodes in the two networks through the four steps in Subsect. 2.3, then the system-level vulnerability V_T from Topology-based viewpoint and V_E from Flow-based viewpoint can be obtained respectively. Obviously, the localized attack strength σ influences the vulnerability results strongly, thus, we investigate the effect of σ on the vulnerability of the whole networks. For illustrative purpose, we set σ range from 0.1 to 5 at increments of 0.1, $\alpha = 0.3$, $\beta = 0.4$, $w_1 = w_2 = 0.5$. Figure 4(a) shows the corresponding result.

From Fig. 4(a), we can see both V_T and V_E increase quickly with increasing σ. It is easy to understand this phenomenon, the bigger σ, the larger affected region caused by the localized attack, and more nodes will be failed, resulting in the larger V_T and V_E. We also find that when σ is big enough, all nodes in the whole networks are failed and both V_T and V_E equal to 1.

Secondly, we identify critical infrastructure components. Similarly, we set each node in both networks as the localized attack center one by one and calculate V_T and V_E respectively. The Fig. 4(b) shows the overall vulnerability of the two networks when 60 nodes in power network are attacked one by one. It can be seen that the nodes {12, 31, 59} are more vulnerable and the nodes {32, 46, 55, 58} are more robust than the others from both Topology-based and Flow-based viewpoints. This is because {12, 31, 59} are located in the center of CIs map and the localized attacks to them can cause more nodes to fail, while {32, 46, 55, 58} are located in the corner, on the contrary. Similar results can be obtained when 30 nodes in water network are attacked one by one, which is shown in Fig. 4(c). The nodes {2, 9, 23} are more vulnerable while {3, 26} are more robust.

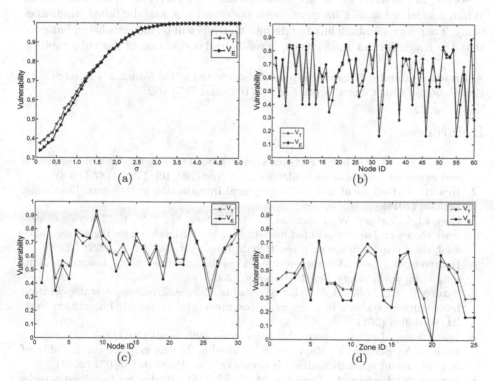

Fig. 4. (a)The effect of localized attack strength σ on the vulnerability of the whole networks; (b)-(c)The vulnerabilities of 60 nodes in power network and 30 nodes in water network respectively; (d) The vulnerabilities of 25 impact zones.

Thirdly, we identify critical geographical locations. We remove 25 impact zones one by one and calculate the corresponding negative consequences. The result is shown in Fig. 4(d). It can be seen that the zones {(2, 1), (3, 2)} are more critical than the others and the zones {(4, 5), (5, 4), (5, 5)} cause smaller negative consequences. Especially for the zone (4, 5), the removal of it does not result in any effect of the overall networks since there is no node located in it.

4   Conclusions

In this paper, we have presented a framework to analyze the vulnerability of CIs under localized attack. Both functional and geographic interdependencies have been taken into consideration in the cascading failures propagation. Localized failure model has been proposed for modelling geographic interdependency and quantifying the failure probabilities of nodes located in the affected region. Multi-layer networks have been employed for establishing functional interdependency. We discussed the vulnerability from the following three perspectives.

Our framework can help stakeholders with increasing their knowledge of the vulnerability of CIs that they are responsible for and better protecting them. However, in the framework, link failure in the affected region is not considered. When a localized attack happens, both node and link may be failed simultaneously. Therefore, combing link failure into analysis will be more realistic, but note that the framework at least gives a lower-bound estimation of the vulnerability.

Acknowledgments. This work is partially supported by the National Natural Science Foundation of China under Grant Nos.71371039 and 71501022.

References

1. Kröger, W.: Critical infrastructures at risk: a need for a new conceptual approach and extended analytical tools. Reliab. Eng. Syst. Saf. **93**, 1781–1787 (2008)
2. Hewitt, K.: Regions of Risk: a Geographical Introduction to Disasters. Routledge, London (2014)
3. Eusgeld, I., Kröger, W., Sansavini, G., Schläpfera, M., Ziob, E.: The role of network theory and object-oriented modeling within a framework for the vulnerability analysis of critical infrastructures. Reliab. Eng. Syst. Saf. **94**, 954–963 (2009)
4. Hellström, T.: Critical infrastructure and systemic vulnerability: towards a planning framework. Safety. Sci. **45**, 415–430 (2007)
5. Faturechi, R., Miller-Hooks, E.: Measuring the performance of transportation infrastructure systems in disasters: a comprehensive review. J. Infrastruct. Syst. **21**, 04014025 (2014)
6. Barthélemy, M.: Spatial networks. Phys. Rep. **499**, 1–101 (2011)
7. Bashan, A., Berezin, Y., Buldyrev, S.V., Havlin, S.: The extreme vulnerability of interdependent spatially embedded networks. Nat. Phys. **9**, 667–672 (2013)
8. Berezin, Y., Bashan, A., Danziger, M.M., Li, D.Q., Havlin, S.: Localized attacks on spatially embedded networks with dependencies. Sci. Rep. **5**, 8934 (2015)
9. Shao, S., Huang, X., Stanley, H.E., Havlin, S.: Percolation of localized attack on complex networks. New J. Phys. **17**, 023049 (2015)
10. Yuan, X., Shao, S., Stanley, H.E., Havlin, S.: How breadth of degree distribution influences network robustness: comparing localized and random attacks. Phys. Rev. E **92**, 032122 (2015)
11. Dong, G.G., Du, R.J., Hao, H.F., Tian, L.X.: Modified localized attack on complex network. Europhys. Lett. **113**, 28002 (2016)
12. Rinaldi, S.M., Peerenboom, J.P., Kelly, T.K.: Identifying, understanding, and analyzing critical infrastructure interdependencies. IEEE Control Syst. **21**, 11–25 (2001)

13. Eusgeld, I., Nan, C., Dietz, S.: "System-of-systems" approach for interdependent critical infrastructures. Reliab. Eng. Syst. Saf. **96**, 679–686 (2011)
14. Marrone, S., Nardone, R., Tedesco, A., et al.: Vulnerability modeling and analysis for critical infrastructure protection applications. Int. J. Crit. Infr. Prot. **6**, 217–227 (2013)
15. Wang, S.L., Hong, L., Chen, X.: Vulnerability analysis of interdependent infrastructure systems: a methodological framework. Physica A **391**, 3323–3335 (2012)
16. Liu, H., Davidson, R.A., Apanasovich, T.V.: Spatial generalized linear mixed models of electric power outages due to hurricanes and ice storms. Reliab. Eng. Syst. Saf. **93**, 897–912 (2008)
17. Wang, J.W., Rong, L.L.: Cascade-based attack vulnerability on the US power grid. Safety Sci. **47**, 1332–1336 (2009)
18. Simonsen, I., Buzna, L., Peters, K., Bornholdt, S., Helbing, D.: Transient dynamics increasing network vulnerability to cascading failures. Phys. Rev. Lett. **100**, 218701 (2008)
19. Ouyang, M.: Review on modeling and simulation of interdependent critical infrastructure systems. Reliab. Eng. Syst. Saf. **121**, 43–60 (2014)
20. Pederson, P., Dudenhoeffer, D., Hartley, S., Permann, M.: Critical Infrastructure Interdependency Modeling: a Survey of US and International Research. Idaho National Laboratory, pp. 1–20 (2006)
21. Buldyrev, S.V., Parshani, R., Paul, G., Stanley, H.E., Havlin, S.: Catastrophic cascade of failures in interdependent networks. Nature **464**, 1025–1028 (2010)
22. Shekhtman, L.M., Danziger, M.M., Havlin, S.: Recent advances on failure and recovery in networks of networks. Chaos, Solitons Fract. **90**, 28–36 (2016)
23. Li, W., Bashan, A., Buldyrev, S.V., Stanley, H.E., Havlin, S.: Cascading failures in interdependent lattice networks: the critical role of the length of dependency links. Phys. Rev. Lett. **108**, 228702 (2012)
24. Shekhtman, L.M., Berezin, Y., Danziger, M.M., Havlin, S.: Robustness of a network formed of spatially embedded networks. Phys. Rev. E **90**, 012809 (2014)
25. Ouyang, M.: Critical location identification and vulnerability analysis of interdependent infrastructure systems under spatially localized attacks. Reliab. Eng. Syst. Saf. **154**, 106–116 (2016)
26. Ouyang, M., Hong, L., Mao, Z.J., Yu, M.H., Fei, Q.: A methodological approach to analyze vulnerability of interdependent infrastructures. Simul. Model. Pract. Theor. **17**, 817–828 (2009)
27. Hong, S., Zhang, X., Zhu, J., Zhao, T.D., Wang, B.Q.: Suppressing failure cascades in interconnected networks: considering capacity allocation pattern and load redistribution. Mod. Phys. Lett. B **30**, 1650049 (2016)
28. Wang, J.W., Rong, L.L., Zhang, L., Zhang, Z.Z.: Attack vulnerability of scale-free networks due to cascading failures. Physica A **387**, 6671–6678 (2008)
29. Haimes, Y.Y.: On the definition of vulnerabilities in measuring risks to infrastructures. Risk Anal. **26**, 293–296 (2006)
30. Johansson, J., Hassel, H.: An approach for modelling interdependent infrastructures in the context of vulnerability analysis. Reliab. Eng. Syst. Saf. **95**, 1335–1344 (2010)
31. Ouyang, M., Pan, Z.Z., Hong, L., Zhao, L.J.: Correlation analysis of different vulnerability metrics on power grids. Physica A **396**, 204–211 (2014)
32. Apostolakis, G.E., Lemon, D.M.: A screening methodology for the identification and ranking of infrastructure vulnerabilities due to terrorism. Risk Anal. **25**, 361–376 (2005)

The Network Topology of the Chinese Creditees

Yingli Wang$^{(\boxtimes)}$, Mingmin Yang, Xiangyin Chen, Changli Zhou,
and Xiaoguang Yang$^{(\boxtimes)}$

Academy of Mathematics and Systems Science, UCAS, Beijing, China
474570979@qq.com, {yangmingmin2005,hvarian}@126.com,
zhouchangli@163.com, xgyang@iss.ac.cn

Abstract. We provide an empirical analysis of network structure of the Chinese creditee market based on a unique data set from the China Banking Regulatory Commission (CBRC). The data set includes guarantee and shareholding relationships among customers of the top 19 commercial banks in China. With these data, we construct three creditee linkage networks: guarantee network, shareholding network and mixed network. Then we employ complex network measures to extract topological characteristics of the three creditee networks. We find that out-degree and in-degree distributions of the three networks are power law and exponential, respectively. In addition, in the three creditee networks, distributions of component size fit power law. Finally, compared with other real networks (such as networks of Facebook, Google+, Twitter, Citation, Paper cooperation, Web link), the average clustering coefficient, connectivity and density of creditee networks are smaller.

Keywords: Complex networks · Creditee networks · Topological characteristic

1 Introduction

The Subprime crisis highlights the important role played by the financial markets in accommodating short-term and long-term fund supply. In November 2008, after the Chinese government released the 4-trillion-stimulus package plan, the amount of the guaranteed loans soared and a number of credit defaults outbroke. The creditee risk spreads through plenty of ways. In this paper, guarantee relationship and the shareholding relationship between the companies are studied as two important channels of creditee risk.

According to the Guarantee Law of China, a guarantor is liable to pay once the debtor defaults on his repayments. So the company who takes a related responsibility guarantee may be asked to pay the debt of its guaranteed company before the bankruptcy of this company. Some companies in the guarantee-chain with debt crisis could impact others on the guaranteed-chain. If one of them confronts problems, it is very likely to have a domino effect. One example of guarantee chain crisis happened in Zhejiang in 2008. Some textile enterprises in Shaoshing (Jianglong Holdings, Hualian Sanxin, Pentacyclic spandex etc.)

© Springer Nature Singapore Pte Ltd. 2016
J. Chen et al. (Eds.): KSS 2016, CCIS 660, pp. 104–114, 2016.
DOI: 10.1007/978-981-10-2857-1_9

bankrupted because of the break of fund chain. A lot of companies guaranteed for them would fall into financial crisis if the bankruptcy comes upon them. The could be worst situation is that the whole economy in Shaoshing fall into crisis one after another on the guarantee chain. Ultimately the local government's rescue holds back the system from being torn down.

Another channel to propagate the creditee risk is through shareholding relationship. If the shareholding enterprises pay too much attention on the investment return brought from the stock prices, it will help to increase the bubble in a bull market. The stock becomes a hot potato in a bear market. The volatility of stock prices in the secondary market brought by cross shareholding has significant effect on the profitability of the company. China Life Insurance and China Ping An Insurance are major shareholders. They hold 20% and 23% companies stocks respectively by the end of 2007. The market values of them are 82.67 billion yuan and 49.1 billion yuan respectively. But when the market began to go through a recession in 2008, they lose much. By 31 March, the market value of China Life Insurance and China Pingan Insurance has decreased by 37% and 32%, respectively. The market value of the companies who hold stocks of China Life Insurance and China Pingan Insurance all decreased by a ratio of more than 50%. So there exists risks of asset bubbles among the shareholding relationships. When the bubbles burst, all the participants will pay a heavy price.

Given all these above, it is quite interesting and meaningful to explore the creditee system in China, especially focus on the real world network of the creditee linkages. In this paper, we employ complex network measures to study Chinese creditee networks. As highlighted by Newman, one prominent advantage of employing network-based theory is that it is able to capture topological and structural characteristics of the data relationships. As a powerful tool for modeling interconnections between economic agents and revisiting the underlying principles of financial fragility, network analysis has been applied by a large body of literature [1,7,9]. Battiston and his coauthors recently confirm the theory of too-central-to-fail by introducing a novel measure of systemic impact inspired by feedback-centrality, and reveal a giant bow-tie structure of the control network of transnational corporations. They also investigate the evolving characteristics of interbank networks under several time scales, and verify that the optimal architecture of financial systems to hold back the default cascading dynamics is problem dependent [2,6,13]. Network analysis based on economic and financial datasets is also investigated in the context of associations between board directors [3], correlations of stock returns [4], interbank market [5], shareholding networks [12], inter-firm networks [11], overnight money market [8], global corporate control [14], and global banking [10].

In this paper, owing to a comprehensive loan-level database, we construct the real creditee networks linked by shareholding relations and guarantee relations, which include guarantee network, shareholding network and the mixed network. The topological characteristics of the three creditee networks are described, and comparisons of these topological characteristics between these creditee networks and some other real networks are made.

Our paper has the following three works. First, we exploit a real-world inter-corporate complex networks with some specific characteristics: in- and out-degree distributions, clustering coefficient, component, average shortest path length. Second, we compare the topological characteristics of Chinese creditee networks with other real networks to have a further understanding of creditee networks. Moreover, due to the scarcity of creditee market data in China, there are few studies in this area, and our research just makes up for the blank in this respect.

The paper proceeds as follows. Section 2 describes the data and network construction. Section 3 presents the characteristics of complex network of Chinese creditee system. Section 4 makes a comparison of the topological characteristics of Chinese creditee networks with other real networks and Sect. 5 concludes this paper by pointing out suggested directions for further research.

2 Data and Network Construction

Our proprietary data set on loan details is provided by China Banking Regulatory Commission (CBRC) which includes two different relationships between enterprises, namely loan guarantee relationship and shareholding relationship. On the one hand, the data set contains the monthly information of all new loans extended to client firms with credit line exceeding 50 million yuan by the 19 national-wide Chinese commercial banks from January 2007 to March 2012. The data set contains company codes of loan borrowing firms and company codes of guarantee enterprises, the amount of money and times the guarantee enterprises provide. On the other hand, our data also cover a cross shareholding relationship from January 2007 to March 2012.

We use three types of creditee networks extracted from the above comprehensive loan-level data: (I) guarantee network. (II) shareholding network. (III) mixed network which integrated the above two networks. In these networks, the node is client firm or related party (i.e., its major stakeholder or guarantee company). In the guarantee network, an oriented edge from node i to node j means that node i has guaranteed a loan for node j. The shareholding network is a directed graph from shareholders to the corresponding corporations. Taking into account the two risk-spreading channels, we merger the above two networks together and name it as mixed network. To get a general understanding of the networks, we snap shot January 2008 from the data set, which has a total number of 112088 nodes, 36734 guarantee edges and 91761 shareholding edges.

3 The Network Topology of Creditees in China

In the following subsections, we analyze the three creditee networks from the following aspects: in-degree and out-degree distributions, component, clustering coefficient, and average shortest path length.

3.1   In-degree and Out-degree Distributions

Degree as a basic parameter, describes structural characteristic of network nodes, and reflects macroscopic statistical property of network system. The degree of node is defined as the number of edges which connect to this node in graph theory. In directed network the degree of node is divided into out-degree and in-degree. Out-degree is the number of edges which is directed from this node to other nodes, in-degree is the number of edges which is directed from other nodes to this node. In directed network the degree of node equals to the sum of out-degree and in-degree.

Firstly, we study the out degree and in degree of three creditee networks: guarantee network, the shareholding network and mixed network. We extract data of January 2008 which reveal degree distributions of the three creditee networks. The distributions of in-degree and out-degree of the networks are shown in Fig. 1. As for other period of the data set, the in- and out-degree distributions are similar.

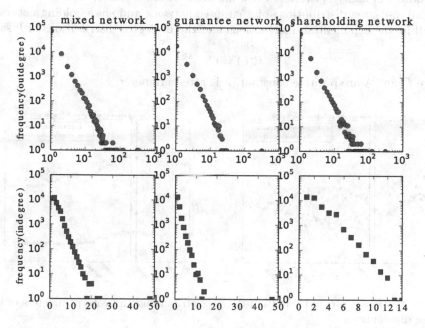

Fig. 1. The in-degree and out-degree distributions of three creditee networks

(I) Figure 1 clearly illustrates that the out-degree distributions of the three networks fit the power law with a fat tail. Power-law distribution implies: (i) Most companies provide guarantees for a very small number of other companies in guarantee network, while they have very limited number of shareholders in shareholding network. (ii) Fat tail indicates for several huge companies, they may either guarantee for quite a few other companies or hold a large amount of other companies' shares.

(II) Compared with the power law distribution of out-degree, the in-degree distributions of three networks are all exponential. The reason for that is a loan contract usually needs to be guaranteed by at most two companies. In shareholding network, though one company may have many shareholders, most of the firms are held by less than 20 firms.

(III) From the upper panels of Fig. 1, the fat tail of the out-degree distribution of the mixed network is mainly because of the shareholding relationship, rather than the guarantee relationship.

It can be concluded that out-degree distributions of guarantee network and shareholding network all fit the power-law distributions and the out-degree distributions of k satisfies:

$$P(k) = Ck^{-\alpha},$$

where C and α are positive constant, α is power law index. In fact, the power law indices of many networks in real life are between 2 and 3.

The in-degree distributions of guarantee network and shareholding network are all fit the exponential distributions and the in-degree distributions of k satisfies:

$$P(k) = Ce^{-\lambda k},$$

where C and λ are positive constant, λ is rate parameter.

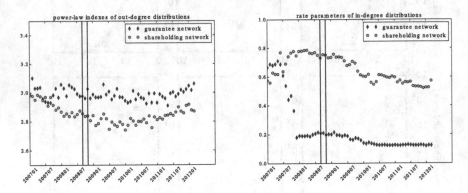

Fig. 2. The power law indices of out-degree distributions and rate parameters of in-degree distributions

We select period from November 2008 to March 2012 and calculate the power law indices of out-degree distributions and rate parameters of in-degree distributions. The results are shown in Fig. 2.

(I) The average power law indices for guarantee network is 2.98, and the average power law indices for shareholding network is 2.82. The results mentioned aboved show high consistence with empirical studies on other scale-free networks.

(II) The average rate parameters of in-degree distributions of the guarantee network and the shareholding network is 0.15 and 0.62, respectively. With the increase of the rate parameter, $P(k)$ decreases in an exponential fashion. Therefore, when k is small, the number of nodes of shareholding networks is smaller compared with guarantee networks. Reversely, when k is large, the number of nodes of guarantee networks is smaller.

(III) Although both the sizes of guarantee network and shareholding network increase rapidly with time, the power law indices and the rate parameters are not significantly changed. It describes that the overall structure of the networks doesn't have significant fluctuation with the rapid growth of the number of nodes. Stable creditee networks are thus clarified.

3.2 Component

The connectivity of the graph plays a critical role in contagion, learning, and the diffusion of various behaviors through a social network. Path relationships in a network naturally partitions a network into different connected subgraphs that are commonly referred to as components. The components of a network are the distinct maximal connected subgraphs of that network.

Figure 3 shows the component size distribution of mixed network, guarantee network and shareholding network for January 2008.

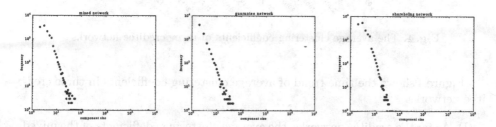

Fig. 3. Component size distributions of three creditee networks

(I) All the three networks have a major component which contains about half of nodes, while the size of other components is less than 100. For example, in the mixed network, The major component which includes 52,001 nodes accounting for 46.39 % nodes of the whole network. The rest of the components' sizes are all less than 100. The guarantee network and the shareholding network also have major components.

(II) The distributions of component size of the three creditee networks perfectly match the power law. However, there is a slight difference between guarantee network and shareholding network. In the shareholding network, the majority components contain 3 nodes. Though in guarantee network, the majority components contain 2 nodes.

3.3   Clustering Coefficient

The clustering coefficient of a node is defined as the proportion of links between the nodes within its neighborhood divided by the number of links that could possibly exist between them. The average clustering coefficients of many real networks have been studied. For example, the average clustering coefficient of World Wide Web is 0.1078, Internet is 0.18-0.3, and Scientific collaboration is 0.4, respectively.

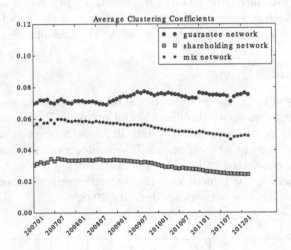

Fig. 4. The average clustering coefficients of three creditee networks

Figure 4 shows the time trend of average clustering coefficients in three creditee networks.

(I)   As for the creditee networks, the average clustering coefficients of the mixed network, guarantee network and shareholding network are 0.0578, 0.0734 and 0.0305, respectively.

(II)   Compared with real networks mentioned above, the clustering coefficients of creditee networks are relatively small and the guarantee relationship is much closer than cross-shareholding relationship.

(III)   The time line of clustering coefficients shows that the closeness of guarantee relationship is stable upward, though closeness of shareholding network is declining.

3.4   Average Shortest Path Length

Average shortest path length describes the separation degree between nodes in a network. Most of real networks have short average path length which is the small-world property. Figure 5 shows the average shortest path length of guarantee network and shareholding network.

(I) Guarantee network and shareholding network exhibit features of small worlds, that is, the ratio of average shortest path to the size of the network is relatively low.

(II) The average shortest path length (about 10) of the shareholding network is shorter than that of guarantee network (about 17), which indicates that cross shareholding relationship is closer compared to guarantee relationship.

(III) From 2009 to 2012, though the scale of networks increased significantly, the average length of the shortest paths remains unchanged.

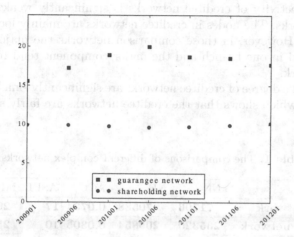

Fig. 5. The average shortest path length of guarantee and shareholding networks

4   The Comparisons of Creditee Networks with Other Complex Networks

In order to discover the unique topological characteristics of creditee networks, and deepen the understanding about the creditee market of China, we will compare the creditee networks with some real networks. We choose social networks, physical networks and web networks as representative of real networks.

The comparisons between creditee networks and other complex networks are listed in Table 1. The networks of Facebook, Google+ and twitter in Table 1 are online social networks which are the interactive network of users. Citation network 1 is the high energy physics Arxiv citation network and citation network 2 is the patent citation network of the United States. Paper cooperation network shows the relations of cooperation between the high Arxiv of energy physics articles. Web link network studies the links between the web pages, data of which is from the Google. In Table 1. NN is short for number of nodes. NE is abbreviated as number of edges. ACC is short for average clustering coefficient.

ASPL is short for average shortest path length. MC is short for major component. AD is short for average degree.

From Table 1, some interesting conclusions are obtained.

(I) From the number of nodes and edges perspective, the scale of creditee networks is much larger than the online social networks.

(II) The most outstanding difference between the creditee networks and those networks is that the average clustering coefficient of creditee networks is very small.

(III) The average shortest paths of all creditee networks are short (compared with the number of nodes) and they all have the properties of small world.

(IV) The connectivity of creditee networks is significantly weaker than other real networks. The nodes in creditee networks are mainly located in small branches. However, in those comparison networks the majority of nodes are located in one branch and the major component tend to be 100 % of the network.

(V) The average degree of creditee networks are significantly smaller than those networks which shows that the creditee networks are fairly sparse.

Table 1. The comparisons of different complex networks

Networks	NN	NE	ACC	ASPL	MP(%)	AD
Guarantee network	**71796**	**70588**	**0.0734**	**17**	**29.2**	**1.82**
Shareholding network	**235320**	**209854**	**0.0305**	**10**	**25.8**	**1.78**
mixed network	**112070**	**246220**	**0.0578**	**7**	**45.3**	**2.52**
Facebook	4039	88234	0.6055	8	100.0	21.85
Google+	107614	13673453	0.4901	6	100.0	127.06
Twitter	81396	1768149	0.5653	7	100.0	21.72
Citation network1	27770	352807	0.3120	13	99.6	24.46
Citation network2	3774768	16518948	0.0757	22	99.7	8.77
Paper cooperation network	12008	118521	0.6115	13	93.3	19.74
Web link network	875713	5105039	0.5143	21	97.7	5.83

5　Conclusion

Based on a comprehensive loan-level database of CBRC, we acquire the characteristics of the complex creditee networks of China which include two kinds of connections, the loan guarantee and the shareholding relationships. We derive three creditee networks from the two different relationships, namely guarantee network, shareholding network and the mixed creditee network. We find that all the out-degree distributions of creditee networks are power law and all the in-degree distributions of creditee networks are exponential. The distributions of component size of creditee networks are power-law. We calculate the average clustering coefficients of the three creditee networks and find that all the networks are very sparse. The average shortest path lengths of guarantee and shareholding network are short which indicates that the guarantee networks and shareholding networks have the property of small-world. By comparing topological characteristics of creditee networks with other social networks, we have reached fundamental understanding about the creditee networks. We find that the average clustering coefficient, connectivity and density of creditee networks are smaller compared with other networks.

Our paper contributes a preliminary step in the study of Chinese creditee networks. There are many intriguing further directions. For instance, it is potentially fruitful to study financial contagion through the creditee linkage networks. It also deserves to study the phenomenon when creditee linkage network is mixed with other channels of contagion, e.g., the monsoonal effect. We are on our way of an in-depth study of Chinese creditee market.

Acknowledgments. This research is supported by the National Natural Science Foundation of China under grant number of '71532013.

References

1. Allen, F., Babus, A.: Networks in Finance. In: Kleindorfer, P., Wind, Y., Gunther, R. (eds.) The Network Challenge: Ctrategy, Profit, and Risk in an Interlinked World, p. 367. Prentice Hall Professional, New York (2009)
2. Battiston, S., Caldarelli, G., Georg, C., et al.: Complex derivatives. Nature Phys. **9**(3), 123–125 (2013)
3. Battiston, S., Catanzaro, M.: Statistical properties of corporate board and director networks. Eur. Phys. J. B-Condens. Matter Complex Syst. **38**(2), 345–352 (2004)
4. Bonanno, G., Caldarelli, G., Lillo, F., Mantegna, R.N.: Topology of correlation-based minimal spanning trees in real and model markets. Phys. Rev. E **68**(4), 046130 (2003)
5. Boss, M., Elsinger, H., Summer, M., et al.: Network topology of the interbank market. Quant. Finan. **4**(6), 677–684 (2004)
6. Delpini, D., Battiston, S., Riccaboni, M., et al.: Evolution of Controllability in Interbank Networks. Sci. Rep. **3** (2013)
7. Goyal, S.: Connections: An Introduction to the Economics of Networks. Princeton University Press, Princeton (2012)
8. Iori, G., Masi, G.D., Precup, O.V., et al.: A network analysis of the Italian overnight money market. J. Econ. Dyn. control **32**(1), 259–278 (2008)
9. Jackson, M.O.: Social and Economic Networks. Princeton University Press, Princeton (2010)
10. Minoiu, C., Reyes, J.A.: Network Analysis of Global Banking: 1978–2009. International Monetary Fund, Washington (2011)
11. Saito, Y.U., Watanabe, T., Iwamura, M.: Do larger firms have more interfirm relationships? Phys. A Stat. Mech. Appl. **383**(1), 158–163 (2007)

12. Souma, W., Fujiwara, Y., Aoyama, H.: Heterogeneous economic networks. In: Namatame, A., Kaizouji, T., Aruka, Y. (eds.) The Complex Networks of Economic Interactions. Lecture Notes in Economics and Mathematical Systems, vol. 567, pp. 79–92. Springer, Heidelberg (2006)
13. Vitali, S., Battiston, S.: Geography versus topology in the European ownership network. New J. Phys. **13**(6), 063021 (2011)
14. Vitali, S., Glattfelder, J.B., Battiston, S.: The network of global corporate control. PloS One **6**(10), e25995 (2011)

Product Diffusion Research Based on Symbolic Regression

Weihua Cui, Xianneng Li, and Guangfei Yang[✉]

Department of Management and Economics,
Dalian University of Technology, Liaoning, China
cuiweihuadlut@163.com,
{gfyang,xianneng}@dlut.edu.cn

Abstract. With the popularity of on-line shopping increasing, on-line products diffusion becomes the growth of importance for marketing decision. Bass diffusion theory is a classical method of forecasting products sales. In this paper, we introduce symbolic regression method to describe the trend of on-line products diffusion and verify whether Bass model in on-line condition performs good as before. Almost all products exhibit seasonality in their sales pattern. Considering the particularity of on-line shopping, we define monthly, weekly and daily period as the division of on-line season. The models perform differently when frequency of data varies.

Keywords: On-line products · Diffusion · Symbolic regression · Seasonality · Bass model

1 Introduction

With the development of online shopping, online products diffusion may perform different features from life-cycle theory. Furthermore, the seasonality based on the classical quarter may confuse consumers and retailers because many products selling on the web are presented no more than 12 months. Online products diffusion and their seasonality with shorter time attract attention of growth of researchers.

The identification and estimation of seasonal influence make contribution to making marketing decisions. Managers will decide the time of introducing new products, how to carry advertisements and the most proper time of raising discount and launching marketing promotion, if the products diffusion is clear. In the previous research, Radas and Shugan mixed seasonal patterns with dynamic models to illustrate their strategical implications [1].

Bass model has been applied to forecast sales when a product is released to the market. Frank M. Bass proposed there were close relation between the number of purchasers and the sale of products, so in his study purchasers were divided into innovators and imitators, which represent the external effect including advertisement without recommendation from others and the internal effect including word-of-mouth produced by previous buyers [2]. Bass also introduced the effects of explanatory variables, such as price and advertisement, to extend the classical model further, which was acted as the Generalized Bass model [3]. Based on both of models, other effort has devoted to new

© Springer Nature Singapore Pte Ltd. 2016
J. Chen et al. (Eds.): KSS 2016, CCIS 660, pp. 115–129, 2016.
DOI: 10.1007/978-981-10-2857-1_10

products diffusion by taking account of network externality, turning point of life-cycle including takeoff and saddle, technology generation and competition [4, 5].

With the life-cycle of products shortening, seasonality plays a significant role of products diffusion. Yuri et, al design dummy variables to capture the sales peak for comparing extended GBM (Generalized Bass model) in which seasonality is modeled with BM (Bass model) which ignores seasonality [8]. There are some arguments about the frequency of data used in Bass model. Time-interval bias exits when we estimate innovation models of new products growth and diffusion with discrete time-series data, such as annual, quarterly and monthly data [6, 7]. The classical method is using X12-ARIMA or TRAMO-SEATS to adjust time series [10]. Yet another method of modeling seasonality is using circular distribution based on nonnegative trigonometric sums developed by Fernandez Duran [9].

Most of products are sensitive to seasonality, for example pharmaceutical drugs, whose sales rise with inflammation breaking out in winter [11]. Sales peak of festival products are usually created before a few days of the festival. If we only consider annual data to forecast products sales as the traditional Bass model does, many important information will be ignored and the precision of forecasting will be challenged. As for electronic products, they are updated quickly and their life-cycles become even less than 1 year, which are too short to estimate the parameters of Bass model, but if monthly or weekly data is available, more items can be used to test and forecast respectively. As on-line shopping popularizes and global goods circulation accelerates, managers cannot wait for estimating the trend of products diffusion at the end of the year. Using short term data to present a whole trend of diffusion becomes noticeable both in academic and practical area. So increasing the frequency of data becomes inevitable in researching on-line products diffusion. A new mechanism should be proposed to adapt to this situation.

The previous researchers use industry data without distinguishing brands to research the products diffusion, such as televisions and CD players [12], or financial reports published by the company, for example iPod [11]. Comparing with offline shopping, on-line shopping has the characteristics of convenience and shortcut, which is growing a fundamental shopping style in life. But it is also suffering many complex external and internal factors which have no influence on the traditional shopping mode. There are a number of results explaining how consumers' behavior impacts on-line products sales and company strategies, but few research about products diffusing on the web. The classical Bass Model describes diffusion rules of durable goods and some modified models emphasize the influence of external factors like seasonality. In terms of on-line shopping, which has complex and volatile shopping environment, the validity of classical Bass model needs to be certified.

Researchers usually assume models of products diffusion based on life-cycle, then use ordinary or generalized least squares method to efficiently optimize parameters. However, we use a data driven method to mine the potential on-line diffusion regulation, without assumed functional forms, verifying whether the Bass model is adaptive to on-line products. Symbolic regression is a classical data-driven method which can effectively discover the optimal parameters, as well as structures of fitting models for the given data. It has been applied in many fields to explore the potential relations or laws. For example, symbolic regression is used to search motion-tracking data automatically [13] and to detect outlier and to extract significant features from the country data [14]. In this paper, this method is firstly applied to discover products diffusion regulation to certify whether on-line products can be described by Bass model or something else. Even another new model could be found to explain on-line products diffusion better than the existed.

Willian [6] and Marielle [7] proposed different frequency of data effected the estimation of parameters in Bass model because of the dispersion degree. The coefficient of innovation with annual data, p, is greater than n p`, where p` is the coefficient of innovation obtained from data of frequency n, whereas the coefficient of imitation with annual data, q, is less than n q`, where q` is the coefficient of imitation with data of frequency n. The Bass model applied in quarterly data performs better than in annual data to forecast the next period sales of products, but does not worse than monthly data. Considering the short life-cycle of on-line products, we use a more intensive frequency of data to explore the basic diffusion. For detecting the effect of seasonality on on-line products diffusion, we compare the results with monthly, weekly and daily data.

For illustrating the research, this article is divided into six parts. Section 1 reviews literature about the effect of seasonality on products diffusion and the hidden regulation about on-line products diffusion. Section 2 introduces the methodology. Section 3 presents the data used in this article and Sect. 4 shows the empirical results. Some discussion about special findings display in Sect. 5. Finally, Sect. 6 concludes findings.

2 Methodology

2.1 Symbolic Regression

Symbolic regression is a function discovery method, used to analyze and model the numeric data set automatically. Without assuming forms of function, this method could generate some mechanism hidden in variables and data set. If using traditional regression method, researchers need to assume the function model with their rich experience and knowledge in the focused field, then Non-linear Least Squares method, Maximum likelihood method or Bayesian estimation are used to estimate the parameters of hypothetical models. However, the range of knowledge of researchers is limited in the age of information explosion and the subjective assumptions of human exist bias. Symbolic regression makes up the gap of discovering models automatically.

In practice, symbolic regression is an evolutionary computation method based on extended genetic programming. In the genetic algorithm, the operator set includes +, − and *, and terminal set contains variables and the number of individuals. The criteria that controls the process of computing could be iteration times or limitations of fitness. The iteration does not stop until reaching the criteria. Replication rate, exchange rate and mutation rate in algorithm are set in advance to control the speed of evolution. Symbolic regression algorithm is a tree structure where every candidate strategy can be found. The genetic operations include reproduction, crossover and mutation with different possibility. Reproduction operation contributes to choose better candidate solution into next generation. Crossover operation gains new traits by resetting parent genes. Mutation operation mutates a node of chosen syntax tree to get a new individual.

Figure 1 presents details of genetic programming. The fitness of models is measured by R-squared (R^2), as follow:

$$R^2 = 1 - \frac{\sum (Y_i - \hat{Y}_i)^2}{\sum (Y_i - \bar{Y}_i)^2}$$

Where Y_i is the actual value, $\hat{Y}_i$ is the predicted value, $\bar{Y}_i$ is the average value.

There usually exist three steps to find proper models with symbolic regression. Firstly, the basic function operators and variables should be selected to express the intrinsic relationship with mathematical models. Secondly, genetic programming is used to evolve model structures and parameters, mainly depending on reproduction, crossover and mutation. Thirdly, we will select the best model to express the relationship based on the quality of models, usually measured by fitness and complexity.

Symbolic regression generates better models with higher precision based on the given data. However, the precise models may fit outlier in the data set. If we apply those models to fit other unknown data, the performance cannot be good as we expected. In other words, with the precision of models increasing, over fitting may appear in the process of finding models. It can be observed that models with higher complexity have higher precision, where over fitting may exit. So the criteria of models need to consider seriously precision and complexity. Some basic methods of evaluating candidate solutions include Akaike information criterion (AIC), Bayesian information criterion (BIC) and Hannan-Quinn criterion (HQ), which depict the loss of the focused model comparing with the real one. But the criterion cannot express the accuracy of the

---

Initialization: Function Set FS, Terminal Set TS, Termination Criterion TC, Max Generation G, Population M, Fitness Measures FM, Reproduction Probability Pr, Crossover Probability Pc, Mutation Probability Pm.

Process:

1. Generate an initial population with FS and TS, gen = 0, and set Pr, Pc, Pm.

2. While gen < G

3.   Calculate fitness of individual using FM

4.   While individual i not satisfy TC:

5.   IF Randomly Probability < Pr

6.   Reproduction: Copy individual i into the next generation

7.   IF Randomly Probability < Pc

8.   Crossover: Recombine randomly individual i and i+1 to create new one into next generation

9.   IF Randomly Probability < Pm

10.   Mutation: Mutate i randomly to create new individual into the next generation

11. gen = gen +1

12. End

Fig. 1. The algorithm of symbolic regression

focused model well. For example, if using such criterion estimating model A, B, C, we find model B performs best, but the validity of model B which explains data set is not guaranteed. In other words, what is possible is that the three models all perform badly, and model B is just relatively better than others. We need to provide some choosing strategies for computers to select accurate models in case of under fitting and over fitting. The method of Pareto Optimality has been used for this case to balance precision and fitness.

2.2   Classical Bass Model

The Bass diffusion is widely used to forecast products sales and technology adoption, as well as to describe the process of interaction between users and potential users which are recognized as innovators and imitators respectively [2, 3]. The basic equation of Bass model as follow:

$$f(t) / (1 - F(t)) = p + q * F(t) \tag{1}$$

where p is the coefficient of innovators, q is the coefficient of imitators, t is the time point, $f(t)$ presents the adoption ratio at time t and $F(t)$ shows the cumulative adoption ratio by time t.

If we know the number of ultimate consumers about the product, usually assumed as m, the sale at time t is $m * f(t)$ and the cumulative sale by time t is $m * F(t)$. So the basic formula for sale as follow:

$$S(t) = m * (1 - F(t)) * (p + q * F(t)) = p * (m - Y(t)) + (q/m) * Y(t) * (m - Y(t)) \tag{2}$$

Where $S(t)$ equaling $m * f(t)$ is the sale at time t and $Y(t)$ equaling $m * F(t)$ is the cumulative by time t. $p * (m - Y(t))$ is the number of innovators and $(q/m) * Y(t) * (m - Y(t))$ is the number of imitators, which represent the adopters affected by external factors like advertisement and the followers affected by internal factors like word-of-mouth respectively.

When we estimate the parameters p, q and m, the following analogous formula could substitute Eq. (2):

$$S(t) = a + b * Y(t) - c * Y(t)2 \tag{3}$$

Where a equals $p * m$, b equals $q - p$ and c equals $-(q/m)$. Finally, the values of p, q and m are calculated from the values of a, b and c.

Based on the Bass model, sales and cumulative sales could be selected as input variables when we use symbolic regression to explore the diffusion of products. If finding the same formula as Eq. (3), we could calculate parameters p, q and m of Bass model in on-line condition. We could also find some particular formulates, differentiating from Bass model, to explore their features further.

Symbolic regression method generates varieties of polymerization based on fundamental operators and data set. The fitness precision and complexity of candidate solutions will be considered thoroughly to select appropriate models to explain online mobile phones diffusion. If the transform of Bass model that is a quadratic polynomial appears in the results of symbolic regression with high performance, we will consider the regression function has similarity with Bass model and classical Bass model can also describe online mobile phones diffusion, but the performance need to be evaluated by new criteria under the condition of online.

3 Data

Differentiating from the previous products adoption study, we choose mobile phones as the research subjects of on-line products diffusion. Electronic products diffusion has been dominant trend gradually in recent years. The speed of electronic products updating is so fast that managers cannot even make appropriate decisions on the basis of their experience synchronously. The academic results that explain the on-line products diffusion provide certain support for making market decisions. Why mobile phones are selected, not cameras, computers et al. The reason is that mobile phones have features of faster speed of updating, more general consumer group and easier to use. Furthermore, in terms of on-line purchasing, mobile phones have shorter life-circle than others, which generates more challenges for managers and deserve researching in the perspective of empirical method.

The resource of data derives from *Jingdong.com* which is the major platform of selling electronic products identified and authorized by official companies. We use parser programming to catch reviews from the web of *Jingdong.com*. The quantity of reviews of the focused mobile phone substitutes real sale. Only after purchasing the product can consumers edit reviews and post on the *Jingdong.com* [18]. Mobile phones have active ranking by sales or rates on the web. If handset vendors stop producing and selling the mobile phone, consumers will not find the product through the original hyperlink. Considering mobile phones updating on the shopping website, we select the products which are released for more than six months and sold by *Jingdong.com* because other third shops have no identification from official companies. Although there are thousands of mobile phones on the *Jingdong.com*, not so many products meet the standard of published period and identification. Finally, we find 76 products for further researching. The statistics information is present in Table 1.

Table 1. Mobile phones statistics information.

Cycle (days)			The number of reviews		
Max	Mean	Min	Max	Mean	Min
891	525.17	180	109427	10326.38	70

4 Results

4.1 Bass Model Appears in on-Line Products Diffusion

If using annual or quarterly sales data of on-line products to explore diffusion, we will lose amount of information and get bias results because of few of items of on-line data.

We select monthly data to verify whether on-line products diffusion has similarity with Bass model. The results are presented in Table 2, in which index of AveofR2 presents the ability of explaining the relationship between F(m) and f(m), and index of Ratio shows the proportion of products covered by the model. The candidate solutions generated by symbolic regression have too many to display, so we rank all models based on the value of ratio. The lower value of ratio means the model cannot fit the samples well and may not use to explain online diffusion properly. So we just present top 20 models in the Table 2. If the candidate solution has similar formula to the Bass model as well as higher value of ratio which can be ranked into top 20, we will admit Bass model can also explain online mobile phones diffusion but which model is the most appropriate will be selected by considering complexity, precision and ratio comprehensively.

Table 2. The filtered models with monthly data.

ID	Model	Complexity	AveofR2	Ratio
M1	f(m) = a * F(m) + b	5	0.2237	0.605
M2	f(m) = −a * F(m) + b	5	0.3537	0.461
M3	f(m) = −a * F(m)^2 + b * F(m) + c	11	0.4988	0.434
M4	f(m) = a * F(m)^3 − b * F(m)^2 + c * F (m) + d	19	0.6498	0.303
M5	f(m) = −a * F(m)^4 + b * F(m)^3 − c * F (m)^2 + d * F(m) + e	29	0.6293	0.303
M6	f(m) = a * F(m)	3	0.2568	0.263
M7	f(m) = −a * F(m)^2 + b * F(m)	11	0.6319	0.237
M8	f(m) = a * F(m)^2 − b * F(m) + c	5	0.5607	0.237
M9	f(m) = −a * F(m)^3 + b * F(m)^2 − c * F (m) + d	19	0.6663	0.197
M10	f(m) = −a * F(m)^3 + b * F(m) + c	13	0.5274	0.184
M11	f(m) = −a * F(m)^2 + b	4	0.3629	0.171
M12	f(m) = −a * F(m)^3 + b * F(m)^2 + c	15	0.6188	0.158
M13	f(m) = −a * F(m)^4 + b * F(m)^3 − c * F (m)^2 + d * F(m)	27	0.6481	0.145
M14	f(m) = a * F(m)^4 − b * F(m)^3 + c * F(m) ^2 − d * F(m) + e	29	0.7050	0.132
M15	f(m) = −a * F(m)^3 + b * F(m)^2 + c * F (m)	23	0.7293	0.118
M16	f(m) = −a * F(m)^4 + b * F(m)^3 − c * F (m) + d	17	0.6067	0.118
M17	f(m) = a * F(m)^4 − b * F(m)^3 + c * F(m) ^2 + d	25	0.6433	0.092
M18	f(m) = −a * F(m)^5 + b * F(m)^4 − c * F (m)^2 + d * F(m) + e	33	0.8361	0.092
M19	f(m) = a * F(m)^3 + b	9	0.4714	0.079
M20	f(m) = a * F(m)^3 − b * F(m) + c	13	0.5888	0.079

Note: In this table, m is the month, $F(m)$ is the cumulative number of reviews by time m and $f(m)$ is the number of reviews on time m. The parameters in the formulas are all positive.

There are four conclusions from Table 2: (1) M3 model has the same format as Bass model, ranking at the third based on Ratio index, which means Bass model still covers 43.42 % mobile phones diffusion and explain the relationship at appropriate 50 % level in terms of on-line condition. (2) M1 and M2 which describe opposite monotonous trend have similar complexity and higher Ratio, but lower average of R2 than other models. Based on the precious research, it means that the product is still primarily developing when quadratic item is zero in the Bass model and only innovated parameter is left. In other words, the product has few followers with the internal effect but a number of innovators with the external effect. (3) M4 and M5 are polynomial with high complexity and strong explanatory ability but can describe only 30 % products. (4) Except for inverted U-shaped (M3 model is similar with Bass model) and monotone curve (M1 and M2 model), we also find N-shaped (M4 model), inverted N-shaped (M9 model) and M-shaped (M5 model) curve appear to describe on-line mobile phones diffusion. In the past decade, many literatures have explored two turning points in the product life cycle: takeoff, which occurs at the beginning, and saddle, which occurs at the early growth [4]. Chandrasekaran and Tellis proposed the saddle in life-cycle curve can be explained through consumer interactions using informational cascade theory [15, 16]. Small shocks to the economic system can temporarily decrease the adoption rate, and the

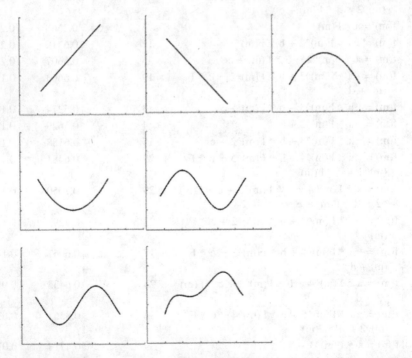

Fig. 2. The main shapes of on-line products diffusion. The abscissas present the cumulative sales by time t and the ordinates present sales at time t.

decrease is magnified through the informational cascade. While heterogeneity in the adopting population is applied to explain this phenomenon. Adopters were divided into two distinct group. If the two groups adopt the innovation at widely differing rates and have little communication, the saddle will appear in sales pattern [17] (Fig. 2).

4.2 Seasonal Effect

On-line products diffusion based on monthly data could not only be described by Bass model, but also explained by other more formulas which may perform better than Bass model in terms of precision or fitness. Seasonality is an important factor impacting products diffusion. The quarterly division does not fit on-line condition synchronously because the average on-line sales cycle is less than two years. In this article, we define weekly and daily effect as seasonality for the fast iteration of on-line products.

The models are selected by Pareto Optimality using weekly and daily data, which are presented in appendix. There exist different models to describe on-line products diffusion based on different frequency of data. Even if a model appears in all three kind of frequency, its precision and fitness perform diversities. Contradict with the moving average method which aims at eliminating the effect of seasonal factors, using reasonable models based on different of frequency of data will provide better support for management decisions.

We also select the common models deriving from different frequency of data, as Table 3. No Bass diffusion model indicates it does not fit any frequency of data under the on-line condition. Those common model curves present monotonous curve, inverted U-shaped, N-shaped, inverted N-shaped and M-shaped, which is consistent with the findings based on monthly data. The diffusion trend of on-line products are similar no matter how the frequency of on-line data is.

Some advice about products diffusion could be proposed according to Table 3. If putting the Ratio to the first place, we can find model L1, L2, L3, L5, L6, L10 fit monthly data better than other two kinds of data, mode L2, L7, L8, L9 fit weekly data

Table 3. The common models based on different frequency of data.

List	Model	Month			Week			Day		
		Ratio	AveofR2	Variance	Ratio	AveofR2	Variance	Ratio	AveofR2	Variance
L1	$f(t) = a * F(t) + b$	0.605	0.224	0.131	0.408	0.151	0.075	0.303	0.103	0.025
L2	$f(t) = -a * F(t) + b$	0.461	0.354	0.145	0.461	0.150	0.053	0.342	0.089	0.030
L3	$f(t) = -a * F(t)^4 + b * F(t)^3 - d * F$ $(t)^2 + e * F(t) + f$	0.303	0.629	0.128	0.316	0.515	0.051	0.105	0.375	0.054
L4	$f(t) = -a * F(t)^2 + b * F(t)$	0.237	0.632	0.094	0.276	0.568	0.045	0.408	0.334	0.048
L5	$f(t) = a * F(t)^2 - b * F(t) + d$	0.237	0.561	0.098	0.184	0.279	0.066	0.132	0.093	0.045
L6	$f(t) = -a * F(t)^3 + b * F(t)^2 - d * F$ $(t) + e$	0.197	0.666	0.091	0.132	0.408	0.127	0.171	0.232	0.073
L7	$f(t) = -a * F(t)^3 + b * F(t) + d$	0.184	0.527	0.167	0.224	0.489	0.040	0.197	0.242	0.045
L8	$f(t) = -a * F(t)^2 + b$	0.171	0.363	0.135	0.250	0.241	0.039	0.211	0.070	0.020
L9	$f(t) = -a * F(t)^3 + b * F(t)^2 + d$	0.158	0.619	0.053	0.197	0.448	0.048	0.158	0.291	0.062
L10	$f(t) = a * F(t)^4 - b * F(t)^3 + d * F(t)$ $^2 - e * F(t) + f$	0.132	0.705	0.048	0.105	0.337	0.145	0.092	0.317	0.040
	Average	0.268	0.528	0.109	0.255	0.359	0.069	0.212	0.215	0.044

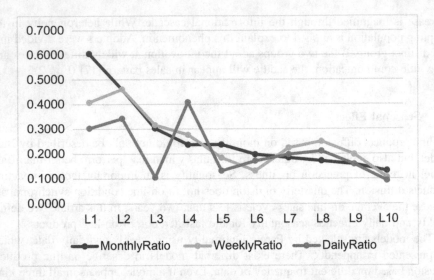

Fig. 3. Comparing Ratio based on different frequency.

better and model L4 fit daily data, as Fig. 3. If the ability of explaining variables relationship is considered prior, all models based on weekly data perform worse than monthly data whose average R2 of ten common models is over 50 %, but better than daily data. More intensive frequency of data may not generate higher precision of diffusion models [6, 7], as Fig. 4. Over concentrated data amplifies the influence of interference when estimating parameters of models.

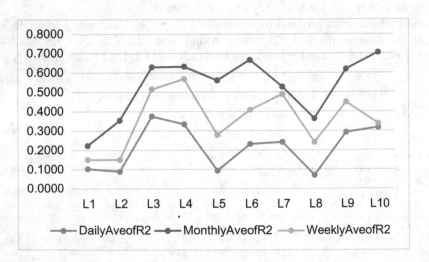

Fig. 4. Comparing Average of R2 of models based on different frequency.

5 Discussion

There are a set of reasonable models returned from symbolic regression for a specific frequency of data with different R-squared value, ratio and complexity. No universal criterion has been proposed as standard measure for model selection in post-processing of symbolic regression. In the precious literatures, some promising criteria have been used to balance error and the number of free parameters, such as AIC, BIC and HQC, which describe the information loss of the focused model. However, the value of error of candidate models is not proportionate to R-squared that presents the precision of models as well as the figure for free parameters is different from the complexity. We propose a new criterion for symbolic regression method to choose the appropriate model, as Eq. (4), which integrates R-squared, ratio and complexity linearly.

$$IRC = \alpha * R_{nor}^2 + \beta * Ratio_{nor} + \lambda * C_{nor} \tag{4}$$

where α, β and λ belong to $[0, 1)$, $\alpha + \beta + \lambda = 1$.

The IRC criterion where α and β are 0.4, λ is 0.2 contributes to select the best model for mobile phones based on the symbolic regression method with monthly data. Table 4 shows models of top 10 IRC values. M4 which presents N-shaped is the best solution to explain online mobile phones diffusion with monthly data. M3 is consistent with Bass model. It can interpret the most products diffusion with the highest value of ratio, but the precision is quite low with the smallest value of average R-squared. So according to the results of symbolic regression, M4 is the most appropriate model to illustrate mobile phones diffusion on the web, not Bass model.

Table 4. The IRC value of models with monthly data.

List	Model	AveofR2	Ratio	Complexity	Value of IRC
M4	f(m) = a * F(m)^3 − b * F(m)^2 + c * F(m) + d	0.650	0.303	29	0.069
M1	f(m) = a * F(m) + b	0.224	0.605	5	0.066
M3	f(m) = −a * F(m)^2 + b * F(m) + c	0.499	0.434	11	0.064
M5	f(m) = −a * F(m)^4 + b * F(m)^3 − c * F(m) ^2 + d * F(m) + e	0.629	0.303	19	0.062
M2	f(m) = −a * F(m) + b	0.354	0.461	5	0.058
M14	f(m) = a * F(m)^4 − b * F(m)^3 + c * F(m) ^2 − d * F(m) + e	0.705	0.132	29	0.056
M18	f(m) = −a * F(m)^5 + b * F(m)^4 − c * F(m) ^2 + d * F(m) + e	0.836	0.092	25	0.054
M9	f(m) = −a * F(m)^3 + b * F(m)^2 − c * F(m) + d	0.666	0.197	19	0.054
M13	f(m) = −a * F(m)^4 + b * F(m)^3 − c * F(m) ^2 + d * F(m)	0.648	0.145	27	0.053
M17	f(m) = a * F(m)^4 − b * F(m)^3 + c * F(m)^2 + d	0.643	0.092	33	0.052

6 Conclusion

On-line products diffusion could be explained by traditional Bass model, but there are many other models fit on-line data better. In this study, we find diversities of diffusion trend through symbolic regression method, such as M-shaped model, N-shaped model and inverted N-shaped model, which have higher precision and fitness than the Bass model. IRC criterion is proposed to select the best model to describe online mobile phones diffusion. For monthly dataset, N-shaped model is the most proper solution to explain online products diffusion. Although Bass model also has quite high value of IRC, its R-squared is too low to interpret the diffusion accurately. Considering seasonality, we find Bass model does not always fit any frequency of data in on-line condition. Even if on-line products is sensitive to seasonal factors, it does not mean the increase of the density of data improves the performance of models. There exist different models properly fitting different frequency of data.

In the further study, we will use the selected model not only to forecast trends of online mobile phones diffusion but to present the sales change in details for supporting market decisions. Also, this subject just focuses on online mobile phones. We believe other items may have similar or different diffusion regulation under the condition of online shopping. If possible, symbolic regression method should be used to verify whether diversities of products on the web have various regression function.

Acknowledgements. This work is supported by the National Natural Science Foundation of China (71671024, 71421001, 71601028), the Fundamental Research Funds for the Central Universities (DUT15RC(3)076), Humanity and Social Science Foundation of Ministry of Education of China (15YJCZH198) and Economic and Social Development Foundation of Liaoning (2016lslktzizzx-01).

A Appendix

See Tables 5 and 6

Table 5. The filtered models with weekly data.

ID	Model	Complexity	AveofR2	Ratio
W1	$f(w) = -a * F(w)^2 + b * F(w) + d$	11	0.431	0.487
W2	$f(w) = -a * F(w) + b$	5	0.150	0.461
W3	$f(w) = a * F(w) + b$	5	0.151	0.408
W4	$f(w) = -a * F(w)^4 + b * F(w)^3 - d * F(w)^2 + e * F(w) + f$	29	0.515	0.316
W5	$f(w) = a * F(w)^3 - b * F(w)^2 + d * F(w) + e$	19	0.547	0.303
W6	$f(w) = -a * F(w)^2 + b * F(w)$	9	0.568	0.276

(*Continued*)

Table 5. (*Continued*)

ID	Model	Complexity	AveofR2	Ratio
W7	f(w) = −a * F(w)^2 + b	7	0.241	0.250
W8	f(w) = −a * F(w)^3 + b * F(w) + d	13	0.489	0.224
W9	f(w) = a * F(w)	3	0.159	0.211
W10	f(w) = −a * F(w)^3 + b * F(w)^2 + d	15	0.448	0.197
W11	f(w) = a * F(w)^2 − b * F(w) + d	11	0.279	0.184
W12	f(w) = a * F(w)^3 − b * F(w)^2 + d * F(w)	17	0.510	0.184
W13	f(w) = −a * F(w)^3 + b * F(w)^2 − d * F(w) + e	19	0.408	0.132
W14	f(w) = −a * F(w)^4 + b * F(w)^3 − d * F(w)^2 + e * F(w)	27.2	0.557	0.132
W15	f(w) = a * F(w)^4 − b * F(w)^3 + d * F(w)^2 − e * F(w) + f	29	0.337	0.105
W16	f(w) = −a * F(w)^4 + b * F(w)^2 + d	17	0.497	0.105
W17	f(w) = a * F(w)^4 − b * F(w)^3 + d * F(w) + e	23	0.613	0.092
W18	f(w) = −a * F(w)^3 + b	9	0.285	0.079
W19	f(w) = a * F(w)^5 − b * F(w)^4 + d * F(w)^3 − e * F(w) + f	35	0.330	0.079
W20	f(w) = −a * F(w)^5 + b * F(w)^4 − d * F(w)^3 + e * F(w) + f	35	0.564	0.079

Table 6. The filtered models with daily data.

ID	Model	Complexity	AveofR2	Ratio
D1	f(d) = −a * F(d)^2 + b * F(d)	9	0.334	0.408
D2	f(d) = −a * F(d) + b	5	0.089	0.342
D3	f(d) = a * F(d) + b	5	0.103	0.303
D4	f(d) = −a * F(d)^2 + b	7	0.070	0.211
D5	f(d) = −a * F(d)^3 + b * F(d) + d	13	0.242	0.197
D6	f(d) = a * F(d)	3	0.074	0.184
D7	f(d) = −a * F(d)^4 + b * F(d)^3 − d * F(d)^2 + e * F(d)	27	0.319	0.184
D8	f(d) = −a * F(d)^3 + b * F(d)^2 − d * F(d) + f	19	0.232	0.171
D9	f(d) = a * F(d)^3 − b * F(d)^2 + d * F(d) + f	19	0.316	0.158
D10	f(d) = −a * F(d)^3 + b * F(d)^2 + d	15	0.291	0.158
D11	f(d) = −a * F(d)^3 + b * F(d)	11	0.367	0.158
D12	f(d) = a * F(d)^3 − b * F(d)^2 + d * F(d)	17	0.256	0.145
D13	f(d) = a * F(d)^2 − b * F(d) + d	11	0.093	0.132
D14	f(d) = −a * F(d)^4 + b * F(d)^3 − d * F(d)^2 + e * F(d) + g	29	0.375	0.105

(*Continued*)

Table 6. (*Continued*)

ID	Model	Complexity	AveofR2	Ratio
D15	$f(d) = -a * F(d)^4 + b * F(d)^3 - d * F(d)^2 + e * F(d) - g$	29	0.240	0.092
D16	$f(d) = a * F(d)^3 - b * F(d)^2 + d$	15	0.189	0.092
D17	$f(d) = a * F(d)^4 - b * F(d)^3 + d * F(d)^2 - e * F(d) + g$	29	0.317	0.092
D18	$f(d) = -a * F(d)^4 + b * F(d)^2 + d$	17	0.360	0.066
D19	$f(d) = -a * F(d)^4 + b * F(d)^3 - d * F(d)^2 + e$	25	0.069	0.066
D20	$f(d) = -a * F(d)^4 + b * F(d) + d$	15	0.279	0.053

References

1. Radas, S., Shugan, S.M.: Seasonal marketing and timing new product introductions. J. Market. Res. **XXXV**, 296–315 (1998)
2. Bass, F.M.: A new product growth for model consumer durables. Manag. Sci. **15**(5), 215–227 (1969)
3. Bass, F.M., Krishnan, T.V., Jain, D.C.: Why the bass model fits without decision variables. Marking Sci. **13**, 203–223 (1994)
4. Peres, R., Muller, E., Mahajan, V.: Innovation diffusion and new product growth models: a critical review and research directions. Int. J. Res. Market. **27**, 91–106 (2010)
5. Bass, F.M., Jain, D., Krishnan, T., et al.: Modeling the marketing-mix influence in new product diffusion. In: Mahajan, V., Muller, E., Wind, Y. (eds.) New Product Diffusion Models. Kluwer Academic Publisher (2000)
6. Putsis, W.P.: Temporal aggregation in diffusion models of first-time purchase: dose choice of frequency matter. Technol. Forecast. Soc. Chang. **51**, 265–279 (1996)
7. Non, M., Franses, P.H., Laheij, C., et al.: Technical note yet another look at temporal aggregation in diffusion models of first-time purchase. Technol. Forecast. Soc. Chang. **70**, 467–471 (2003)
8. Peers, Y., Fok, D., Franses, P.H.: Modeling seasonality in new product diffusion. Market. Sci. **31**(2), 351–364 (2013)
9. Fernandez-Duran, J.J.: Circular distributions based on nonnegative trigonometric sums. Biometrics **60**, 499–503 (2004)
10. Fernandez-Duran, J.J.: Modeling seasonal effects in the bass forecasting diffusion model. Technol. Forecast. Soc. Chang. **88**, 251–264 (2014)
11. Guidolin, M., Guseo, R.: Modelling seasonality in innovation diffusion. Technol. Forecast. Soc. Chang. **86**, 33–40 (2014)
12. Christophe, V.D.B.: New product diffusion acceleration: measurement and analysis. Market. Sci. **19**(4), 366–380 (2000)
13. Schmidt, M., Lipson, H.: Distilling free-form natural laws from experimental data. Science **324**(5923), 81–85 (2009)
14. Kotanchek, M.E., Vladislavleva, E.Y., Smits, G.F.: Symbolic regression via genetic programming as a discovery engine: insights on outliers and prototypes. In: Riolo, R., O'Reilly, U.-M., McConaghy, T. (eds.) Genetic Programming Theory and Practice VII, pp. 55–72. Springer, US (2010)

15. Golder, P.N., Tellis, G.J.: Growing, growing, gone: cascades, diffusion, and turning points in the product life cycle. Market. Sci. **23**(2), 207–218 (2004)
16. Chandrasekaran, D., Tellis, G.J.: Getting a Grip on the Saddle: Cycles, Chasms, or Cascades? PDMA Research Forum, Atlanta, 21–22 October 2006
17. Bulte, C.V.D., Joshi, Y.V.: New product diffusion with influentials and imitators. Market. Sci. **26**(3), 400–421 (2007)
18. Lu, Q., Ye, Q., Law, R.: Moderating effects of product heterogeneity between on-line word-of-mouth and hotel sales. J. Electron. Commence Res. **15**(1), 1–12 (2014)

Public Policy Simulation Based on Online Social Network: Case Study of Chinese Circuit Breaker Mechanism

Yuan Huang[1,2], Yijun Liu[2], and Qianqian Li[2(✉)]

[1] University of Chinese Academy of Sciences, Beijing 100190, China
hnhuangyuan@163.com
[2] Institute of Policy and Management, Chinese Academy of Sciences,
Beijing 100190, China
{yijunliu,lqqcindy}@casipm.ac.cn

Abstract. This paper presents a public policy simulation method established based on the structure and function of online social networks. The simulation technique draws on Markov switching methodology and includes data collection and parameter extraction. Two parameters are crucial to operating the simulation: the initial attitude matrix and the attitude transition probability matrix. Simulation results are obtained via iterative operation of these matrices until reaching equilibrium. This kind of processing method provides a good way to combine the simulation analysis and empirical situation. A case study on the hotly debated "circuit breaker mechanism" policy was conducted to verify the effectiveness of the proposed method. The simulation results suggested that the circuit breaker mechanism is infeasible; the fact that the policy was indeed formally terminated confirms the effectiveness of the simulation method.

Keywords: Public policy simulation · Sentiment classification · Markov switching method

1 Introduction

Policy decision-making is an increasingly massive, complex, and dynamic endeavor related to the large scale and ever-changing conditions to which policies must be applied, such as in marketing [1], environmental protection [2], management [3], and carbon trading [4] fields. The fact that new policies are often necessitated within very brief time periods makes these decisions even more challenging to make. New utilities and methodologies are necessary and urgently to allow policy decisions to be made in the most scientific manner possible. Simulation is a powerful tool for understanding and assessing the operation of systems [5], and are likely the most effective and accurate way to study policy-related issues. Public policy is the basis of effective societal operation and has a significant impact on our daily lives. Improper or ineffective public policies

© Springer Nature Singapore Pte Ltd. 2016
J. Chen et al. (Eds.): KSS 2016, CCIS 660, pp. 130–139, 2016.
DOI: 10.1007/978-981-10-2857-1_11

not only impact human quality of life, but can harm government credibility. Policy-making must be rational, scientific, and effectual, and simulating policies appropriately can help to make this so.

When new public policies are announced, they often cause heated debate both in real-world interpersonal interactions and online. Public attitudes towards a given policy reflect its efficacy: If the public supports the policy, they express positive opinions and vice versa. Public attitudes about policies expressed in the real world can be difficult to obtain, but opinions expressed online (e.g., through social network platforms) can be gathered fairly readily via data capture technology.

There are several techniques available for simulating public policy, such as the agent-based model or system dynamics model. The parameters necessary to run these models are difficult to define, however. In this study, we extracted simulation parameters (the initial attitude distribution and a transition probability matrix) from an online social network and used Markov switching to simulate the public's assessment of a given policy. As opposed to other simulation methods, the method proposed here prevents arbitrary parameter setting. As discussed below, we used a case study to prove the efficiency and practical significance of the proposed method.

2   Markov Switching Method

The Markov decision method can be utilized to predict the possible states of objects in the future based on the current states and variation tendencies of parameters as-observed in the present. Predictions and decisions can be made accordingly by using a transition probability matrix. Markov switching has been used to successfully predict GDP growth [6], traffic safety [7], policy bandwagons [8], exchange rates [9], and simulate energy-water-climate change [10], clinical decision-making [11].

Generally speaking, the transition probability matrix is applied to decision-making as a four-step process: (1) Building the matrix, (2) creating a forecast accordingly, (3) calculating and obtaining the equilibrium state of the matrix, and (4) reading the matrix appropriately in order to make an optimal decision.

2.1   Transition Probability Matrix

The transition probability matrix of the Markov model is:

$$P^{(k)} = \begin{bmatrix} p_{11}^k & p_{12}^k & \cdots & p_{1n}^k \\ p_{21}^k & p_{22}^k & \cdots & p_{2n}^k \\ \vdots & \vdots & \vdots & \vdots \\ p_{m1}^k & p_{m2}^k & \cdots & p_{mn}^k \end{bmatrix}$$

where p_{ij} is the probability value. Transfer probability matrix $P^{(k)}$ has two crucial characteristics: First, that all the elements in the matrix are non-negative, that is, $p_{ij} \geq 0$; and second, that the sum of each row in the matrix is 1, that is, $\sum_{j=1}^{n} p_{ij} = 1$. $k = 1$ is the first step of applying the transition probability matrix: When $k > 1$, Eqs. (1) and (2) are valid [12]:

$$P^{(k)} = P^{(k-1)}P \qquad (1)$$

$$P^{(k)} = P^k \qquad (2)$$

The elements of the transfer probability matrix are derived according to the flow of gains and losses of the research object during certain time periods. The probability is only related to the previous observation of probability and is independent of earlier observations of said probability.

2.2   Equilibrium State of Transition Probability Matrix

The iterative process is complete once the system reaches equilibrium (i.e., there is no longer consistent change in the probabilities observed.) The relationship among probability observations can be represented by Eq. (3); Eq. (4) represents the state of equilibrium.

$$\theta(n+1) = \theta(n)P \qquad (3)$$

$$\theta(n+1) = \theta(n) \qquad (4)$$

3   Parameters Provided by the Online Social Network

For the purposes of this study, we obtained the initial attitude distribution and the attitude transition probability matrix from an online social network. The detailed simulation process is described below (Fig. 1).

3.1   Data Collection

Data crawls from Sina Weibo were collected with Octoparse (a software developed recently). The pseudocode of the data extraction process is provided below.

---

1. Set $T1$ as start time when officials solicit ideas from public;
   set $T2$ as end time of advice-seeking.
2. Prepare dataset.
   2.1 Dataset-1 to construct initial matrix:
       Crawling online information from $T1$ to $T1 + M$ days with specific policy topic.
   2.2 Dataset-2 to construct transformation matrix:
       Crawling online information from $T1 + M$ days to $T2$ with N hashtags about specific policy topic.(M is a variable, $0 < M < T2 - T1$, depending on specific occurrences.)

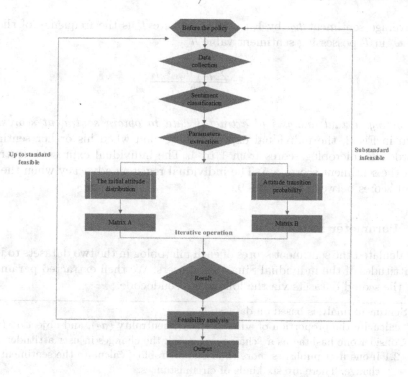

Fig. 1. Public policy simulation based on online social network flow chart

3.2   Sentiment Classification

We applied sentiment classification to define individual attitudes regarding certain events. This included several steps.

Creating the sentiment lexicon. First, we extracted nine levels of sentiment ranging from negative to positive.

Second, we filtered the sentimental stem based on subjective emotion classification; we then collected 1,200 high-frequency words occurring in Weibo directly expressing emotions and 50 buzzwords likewise expressing emotions.

Third, we scored these individual words on a scale of 1 (negative) to 9 (positive) resulting in a measure of average sentiment for each given word. We used h_{ω_i} to signify the sentiment score of the word. For example, "perfect" is one of the most positive words in the list with a score of $h_{\omega_i} = 9$, while "corruption" is one of the most negative at $h_{\omega_i} = 1$. Neutral words like "curiosity" tend to score in the middle of the scale, i.e., $h_{\omega_i} = 5$.

Conducting textual analysis of extracted data to obtain sentiment scores. We used a Natural Language Toolkit named jieba to slice, then match words with the sentiment lexicon. For a given text T containing N unique words, we calculated the average sentiment h_T by Eq. (5) [13]. where f_i is the frequency of the ith word ω_i in T possessing sentiment value h_{ω_i} :

$$h_T = \frac{\sum_{i=1}^{N} h_{\omega_i} * f_i}{\sum_{i=1}^{N} f_i} \tag{5}$$

Conducting textual analysis of extracted data to obtain sentiment scores. As shown in Fig. 2, the individual expresses objection when his or her sentiment towards the microblog scores from 1 to 4. The individual expresses neutrality when the sentiment scores a 5. The individual registers advocacy when the sentiment scores between a 6 and a 9.

3.3 Parameter Extraction

We calculated the sentiment scores of each microblog in the two datasets to judge the attitudes of the individual Sina Weibo users. We then extracted parameters from the scored datasets via the following pseudocode.

---

1. Sentiment analysis based on dataset-1;
    calculate the proportion of advocacy (n_1), neutrality (n_2), and objection (n_3).
2. Consider one hashtag as a "chatroom", track the changes in user attitude:
    2.1 If one user publishes more than one microblog, calculate the sentiment change. There are six kinds of circumstances:
        advocacy -> neutrality
        advocacy -> objection
        neutrality -> advocacy
        neutrality -> objection
        objection -> advocacy
        objection -> neutrality
    We use x_{11}, x_{12}, x_{21}, x_{22}, x_{31}, x_{32} to denote the proportion of the change in all users.
    2.2 If one user publishes one microblog,
    we assume that user's attitude remains unchanged.
    2.3 Obtain transition probability matrix

$$
\begin{array}{c}
\\
advocacy \\
neutrality \\
objection
\end{array}
\begin{array}{ccc}
advocacy & neutrality & objection \\
\left(\dfrac{n_1-x_{11}-x_{12}}{n_1}\right. & \dfrac{x_{11}}{n_1} & \dfrac{x_{12}}{n_1} \\
\dfrac{x_{21}}{n_2} & \dfrac{n_2-x_{21}-x_{22}}{n_2} & \dfrac{x_{22}}{n_2} \\
\dfrac{x_{31}}{n_3} & \dfrac{x_{32}}{n_3} & \left.\dfrac{n_3-x_{31}-x_{32}}{n_3}\right)
\end{array}
$$

4   Case Study: Circuit Breaker Mechanism

We used a well-publicized "circuit breaker mechanism" policy as a case study to assess the feasibility and effectiveness of the proposed policy simulation method.

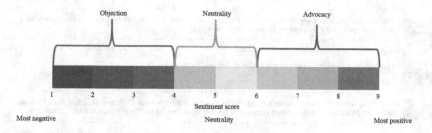

Fig. 2. Attitude judgements of individuals

4.1   Case Summary

The circuit breaker mechanism policy was one of the most fervently discussed public policies enacted in China between September 2015 and January 2016. A timeline of the policy is shown in Fig. 3. The circuit breaker mechanism is a form of protection first established by the US Securities and Exchange Commission. The CSI 300 is the benchmark index for China's circuit breaker mechanism. If the 5 % threshold is triggered during a late period (from 14:45 until 15:00) or if the 7 % threshold is triggered any time during a trading day, trading is suspended until the day's closing. Figure 3 shows the timeline of the Chinese circuit breaker mechanism's operation.

4.2   Results

Data Collection. We used crawling online information about circuit breakers from September 7, 2015 to September 11, 2015 as an initial dataset. We then built the transition dataset based on crawling microblogs hashtagged "circuit breakers mechanism" and "circuit breakers" from September 12, 2015 to September 21, 2015.

Parameter Extraction. The detailed process through which we textually analyzed the sentiments expressed in the obtained datasets is specified in Sect. 3.2. After completing this process, we extracted initial attitude matrix A from the initial dataset and the attitude transition probability matrix B from the transition dataset. Attitude transition probability is depicted in Fig. 4. The end result of matrix A and B is as follows:

$$A = \begin{bmatrix} 0.7 & 0.2 & 0.1 \end{bmatrix}$$

$$B = \begin{bmatrix} 0.4 & 0.3 & 0.3 \\ 0.2 & 0.4 & 0.4 \\ 0.1 & 0.3 & 0.6 \end{bmatrix}$$

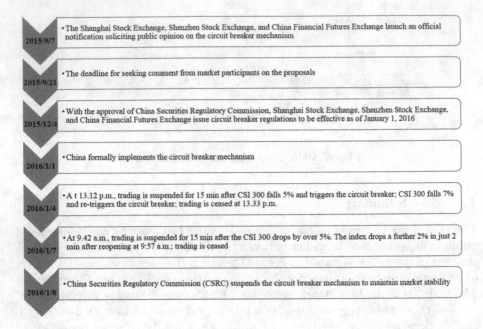

Fig. 3. Circuit breaker policy event timeline

Results. Based on Markov switching method and matrix A and B, we obtained the results shown in Table 1 through iterative computations until reaching equilibrium.

4.3   Feasibility Analysis

We found that individual attitudes toward circuit breakers reached equilibrium (i.e., stability) at the sixth policy simulation step (Fig. 5).

The proportion of "non-objection" (advocacy and neutrality) attitudes expressed towards a given policy reflects the public's approval of the policy. Policy feasibility standards call for a proportion non-objection greater to 0.75, at which point the policy can be considered credible. When the proportion of non-objection is greater than 0.85, the policy has "high-quality credibility"; when the proportion of non-objection is greater than 0.95, the policy has "safe credibility" [14].

As shown in Fig. 6, the proportion of non-objection was calculated to be 0.524 at the equilibrium state. According to policy feasibility standards, the circuit breaker mechanism was thus not feasible.

Our policy simulation results, as discussed above, implied that the circuit breaker mechanism is likely to fail. In actuality, the circuit breaker mechanism was placed into effect on January 1, 2016 and officially suspended on January 8, 2016. If appropriate policy simulation had been conducted prior to January 1,

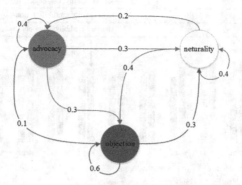

Fig. 4. Attitude transition probability

Table 1. Simulation results

Simulation step	Advocacy	Neutrality	Objection
step0	0.700	0.200	0.100
step1	0.330	0.320	0.350
step2	0.231	0.332	0.437
step3	0.203	0.333	0.464
step4	0.194	0.333	0.473
step5	0.192	0.333	0.475
step6	0.191	0.333	0.476
step7	0.191	0.333	0.476
step8	0.191	0.333	0.476
step9	0.190	0.333	0.476

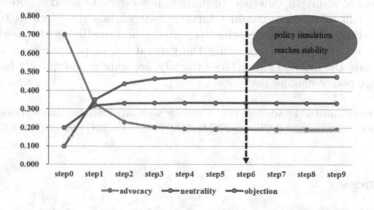

Fig. 5. Policy simulation of circuit breakers reaches stability

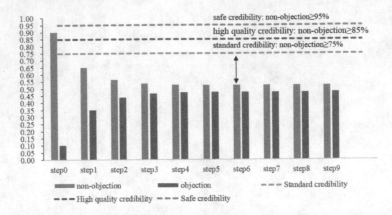

Fig. 6. Circuit breaker mechanism fails to reach expected standards

the erroneous policy implementation could have been prevented (or at least, the policy could have been ameliorated in time to minimize any negative impact.)

5   Conclusion

As discussed above, we utilized a case study to prove the credibility of a new simulation method that we developed based on social network data collection and Markov switching. All the data necessary to establish parameters appropriately are readily available and verifiably relevant, unlike other similar simulation methods in which parameters may be set arbitrarily. Data collection is the basis and parameter extraction is the key to the proposed method, so it is necessary to crawl the valid data and build the updated sentiment lexicon in real-time. Only then can the accuracy and credibility of the parameters be ensured, which enhances the predictability of the policy being simulated.

We plan to apply this method to more policy events to build a policy event store that will allow us to conduct further in-depth analysis, such as by verifying the effectiveness of the method repeatedly and modifying the method as necessary; we also plan to generalize the general law of policy events accordingly to make the method able to scientifically, accurately, and quickly facilitate appropriate policy-making decisions.

Acknowledgements. Funds for this research was provided by the National Natural Science Foundation of China (NSFC)71403262; 71573247; 91024010; 91324009; 71503246.

References

1. Bronnenberg, B.J., Rossi, P.E., Vilcassim, N.J.: Structural modeling and policy simulation. J. Mark. Res. **42**, 22–26 (2005). doi:10.1509/jmkr.42.1.22.56887

2. Parks, P.J., Kramer, R.A.: A policy simulation of the wetlands reserve program. J. Environ. Econ. Manag. **28**, 223–240 (1995). doi:10.1006/jeem.1995.1015

3. Heshmati, A., Lenz, V.F.C.: Policy simulation of firms cooperation in innovation. Res. Eval. **24**(3), 293 (2015)

4. Yuan, Y., Na, L., Shi, M., (2015) A multi-regional CGE model, its application in low carbon policy simulation in China. In: IEEE/Wic/ACM International Conference on Web Intelligence and Intelligent Agent Technology, pp. 36–39. IEEE press, New York, 2015. 10.1109/WI-IAT.2015.18

5. Leros, A., Andreatos, A.: Using Xcos as a teaching tool in a simulation course. In: 6th International Conference on Communications and Information Technology, and Proceedings of the 3rd World Conference on Education and Educational Technologies, pp. 121–126. WSEAS, Wisconsin (2012)

6. Buckle, R.A., Haugh, D., Thomson, P.: Markov switching models for GDP growth in a small open economy. J. Bus. Cycle Meas. Anal. **2004**, 13 (2004). doi:10.1787/17293626

7. Malyshkina, N.V., Mannering, F.L., Tarko, A.P.: Markov switching negative binomial models: an application to vehicle accident frequencies. Accid. Anal. Prev. **41**, 217–226 (2009). doi:10.1016/j.aap.2008.11.001

8. Explaining Policy Bandwagons with Markov models. http://ecpr.eu/Filestore/PaperProposal/93926bc9-a1fb-447b-8cd9-2c46894a79af.pdf

9. Nikolsko-Rzhevskyy, A., Prodan, R.: Markov switching and exchange rate predictability. Int. J. Forecast. **28**, 353–365 (2012). doi:10.1016/j.ijforecast.2011.04.007

10. Nanduri, V., Saavedra-Antolnez, I.: A competitive Markov decision process model for the energy-water-climate change nexus. Appl. Energy **111**, 186–198 (2013). doi:10.1016/j.apenergy.2013.04.033

11. Bennett, C.C., Hauser, K.: Artificial intelligence framework for simulating clinical decision-making: a Markov decision process approach. Artif. Intell. Med. **57**, 9–19 (2013). doi:10.1016/j.artmed.2012.12.003

12. Liang, Z.L., Jia, X.F.: Theory and Practice of Decision Support System. Tsinghua University Press, Beijing (2014)

13. Mitchell, L., Frank, M.R., Harris, K.D., et al.: The geography of happiness: connecting twitter sentiment and expression, demographics, and objective characteristics of place. Plos One **8**, e64417 (2013). doi:10.1371/journal.pone.0064417

14. Niu, W.Y.: Three principles of optimizing social governance structure. Bull. Chin. Acad. Sci. **30**, 61–70 (2015). doi:10.16418/j.issn.1000-3045.2015.01.009

The Online Debate Networks Analysis:
A Case Study of Debates at Tianya Forum

Can Wang and Xijin Tang(✉)

Institute of Systems Science, Academy of Mathematics and Systems Science,
Chinese Academy of Sciences, Beijing 100190, People's Republic of China
wangcan@amss.ac.cn, xjtang@iss.ac.cn

Abstract. In this study, we examine the characteristics of online debate networks. We empirically investigate the debate networks formed by three hot threads from Tianya Forum at individual, whole-network and triad levels. At the individual level, the statistical analysis reveals that people participate different threads about one issue; authors reply to themselves; the authors of the original posts are the core of the interaction; we rank the indegree value and betweenness value of the authors, and find that they are not consistent in sequence. At the whole-network level, the structural indices reveal that the stances of the original posts affect the debate networks. At the triad level, the proportions of coded triads reveal that the common forms in debate networks are mutual dyads; the proportions of triadic closures reveal that relations between participants are different in the two camps; and the balanced triads between camps are more than those within camps.

Keywords: Debate networks · Online social network · Triads · Tianya Forum

1 Introduction

We are in an era where people can easily voice and exchange their opinions on the Internet through social media such as online forums. It is widely recognized that mining public opinion from on-line discussions is an important task, especially in the modern democratic life. Polarization phenomena often happen within the public discussions because that people have different cognitions towards one thing. There exist two streams of literature in this domain. One is automatically determining the debate participants' opinions by text mining [1–3]. The other is detecting the behavior patterns during the discussions [4]. Social network analysis can reveal the interactions patterns of the discussion participants.

The debate networks are different from other online social networks because the dyadic ties between social actors are special. An author may support or disapprove of another one. Some existing literatures focused on the debate networks [4, 5]. Most of the existing researches investigated the debate networks at the whole-network level [5]. In this paper, we analyze a classical debate topic in China at the individual, whole-network and triad levels.

© Springer Nature Singapore Pte Ltd. 2016
J. Chen et al. (Eds.): KSS 2016, CCIS 660, pp. 140–150, 2016.
DOI: 10.1007/978-981-10-2857-1_12

2 Reply Networks and Data Sets

In this section, we define the construction of reply networks at forums and introduce the research data.

2.1 Introduction to Reply Networks

Online forums contain rich threaded discussions (threads) on all kinds of topics/issues, e.g., technology, sports, religion, and politics. In forums, a discussion starts when a user posts an initial post initiating a conversation in a particular matter. Afterwards, other users reply to the initial post or to another reply. The initial post and these replies form a thread.

In this paper, the relationships of the users' reply in a thread give rise to an reply network. The vertices (nodes) are authors and the links represent an author comments on another author's previous messages. The network is directed and there are at most two links between author i and author j ($i \rightarrow j$, $j \rightarrow i$). In the real world, two users may interact for many times but we do not count the weights of the links in this paper.

2.2 Debates on TCM at Tianya Forum

In China, there exist two camps of people according to their attitudes towards traditional Chinese medicine (TCM). Some people take the "abolishing TCM" stance that TCM should be abolished from the national health system. The other camp of people take the "preserving TCM" stance and insist that TCM should be preserved. In our previous research, we noticed that the discussions about TCM were always polarized and we did text analysis to mine the diverse opinions about TCM [3]. In this paper, we choose the classical controversial issue to mine people's interaction patterns among debates.

Tianya Forum is one of the most popular Chinese BBS sites. Table 1 lists three hottest threads about TCM at Tianya Forum. In this paper, we analyze online reply networks formed by the three threads. Reply network for Thread 1 (Network 1) has 4890 authors. Reply network for Thread 2 (Network 2) has 5514 authors. Reply network for Thread 3 (Network 3) has 6065 authors.

Table 1. Hot threads about TCM at Tianya Forum [6]

Thread ID	Replies	Participants	Start time	End time
1	117318	4890	2012-10-16	2013-11-29
2	36592	5514	2011-03-21	2015-01-24
3	33547	6065	2011-11-12	2015-01-24

3 Descriptive Statistics

In this section, we examine the characteristics of online reply networks at the individual level. Descriptive statistics of the participants' activities are listed.

3.1   Self Reply

Within the three threaded discussions, people can reply to posts to state their opinions. Interestingly we find that people sometimes reply to their own posts to strengthen their opinions or to make sure that their opinions are noticed. Table 2 lists the number of authors who reply to themselves. The statistics of the replies which are comments of some ones' own posts are also listed in Table 2.

Table 2. Authors reply to themselves and the corresponding replies

Tread ID	Authors		Replies	
	Size	Percentage	Size	Percentage
1	288	5.88 %	20760	17.70 %
2	191	3.46 %	1602	4.38 %
3	164	2.70 %	1428	4.26 %

3.2   Participate in Different Threads

Thread 1, Thread 2 and Thread 3 are all about the issue of TCM. By comparing the participants we find that some authors reply to different threads about one issue. As shown in Fig. 1, 364 authors participate both in Thread 1 and Thread 3, 174 authors participate both in Thread 1 and Thread 2, 261 authors participate both in Thread 2 and Thread 3, and 92 authors participate in the three threads.

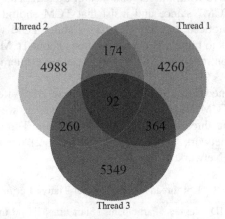

Fig. 1. The venn diagram of the authors in the three threads

3.3   Consistent Relationships Across Networks

In the three networks, we find same reply relationships among authors which means that some of the linking relationships among these authors are consistent across networks. For example, there are 14 pairs of direct links exist in Network 1 and Network 2, 80 pairs of direct links exist in Network 1 and Network 3, 23 pairs of direct links exist in

Table 3. Examples of consistent relationships

Source (author ID)	Target (author ID)	Thread 1	Thread 2	Thread 3
50425395	15956478	17	2	48
11440453	50425395	8	45	21
50425395	13010103	2	1	5

Network 2 and Network 3. Table 3 lists some examples of consistent relationships and the appearing times of these links in the threads.

3.4 Key Players

Key players in networks are determined by their topology attributes and structural attributes. Indicators of centrality identify the most important vertices within a graph. In this paper, we use indegree centrality and betweenness centrality to identify the key players. Table 4 lists the top 10 authors by indegree value. Authors of the original posts ("61681904", "3865013" and "60219641") gain the highest indegree. In other words, the authors of the original posts are the core of the participants.

Table 4. Top 10 authors in decreasing order of their indegree values

Thread 1		Thread 2		Thread 3	
Author ID	Indegree	Author ID	Indegree	Author ID	Indegree
61681904	854	3865013	779	60219641	600
60233507	198	34658621	300	50425395	217
68378554	125	50425395	275	60480958	157
73117788	108	26546902	168	42004025	90
61908805	99	35721022	164	61908805	83
56462278	92	**37971276**	**159**	60244454	72
50425395	91	**5695593**	**154**	15956478	68
72458058	88	53167987	141	2558039	52
47548691	**84**	**14734994**	**138**	**13249554**	**46**
64925795	**64**	**34888324**	**120**	**48969130**	**43**

Betweenness centrality is an indicator of a node's centrality in a network. The betweenness value of a node equals to the number of shortest paths from all vertices to all others that pass through that node[1]. Table 5 lists the top 10 authors by their betweenness value. The bold characters in Table 4 are these authors who do not appear in Table 5. The bold characters in Table 5 are these authors who do not appear in Table 4. We can infer that the indegree value and betweenness value of the authors are not consistent in sequence.

---

[1] https://en.wikipedia.org/wiki/Betweenness_centrality.

Table 5. Top 10 authors in decreasing order of their betweenness values

Thread 1		Thread 2		Thread 3	
Author ID	Betweenness	Author ID	Betweenness	Author ID	Betweenness
61681904	2074041	3865013	713402	50425395	2211015
60233507	317674	26546902	201356	**50191233**	**1965446**
61908805	166961	50425395	146851	60219641	1896792
50425395	147006	34658621	143833	60480958	1329253
73117788	133075	34306001	86521	42004025	265711
72458058	113751	39086442	60060	61908805	259360
74186058	**106932**	53167987	55408	15956478	227940
39086442	**92645**	**13561653**	**53886**	60244454	201634
68378554	88941	35721022	49680	**28721827**	**155379**
56462278	88827	**5608751**	**48757**	2558039	132718

4   Structural Characteristics of the Reply Networks

Structural indices such as connected components, average degrees, clustering coefficients, etc. are given in this part to reveal the structural characteristics of the reply networks at the whole-network level.

4.1   Connected Components

Table 6 lists the sizes of the authors of Thread 1, Thread 2 and Thread 3. In Table 6, authors interact with original post authors are whose who only reply to the original posts. Authors interact with others refer to authors having links with others because that they reply to others' replies. C is the number of connected components in Table 6, GC is the sub-graph with most vertices, and the number of nodes in GC is |GC|.

Table 6. Connected components and GC

Thread ID	Size of authors	Authors interact with original post authors	Authors interact with others	C	|GC|
1	4890	2132	2758	15	2737
2	5514	3348	2166	39	2100
3	6065	1858	4207	9	4193

2758 authors in Network 1 interact with other authors and they make up 56.40 % of the participants of Thread 1. In Network 2, the authors who interact with other authors make up 39.28 % of the participants. In Network 3, the authors who interact with other authors make up 69.37 % of the participants.

4.2 Average Degree

Degree of a vertex of a graph is the number of edges incident to the vertex[2]. Average degree in a network with N nodes is $\bar{d}_i = \frac{\sum_{i=1}^{N} d_i}{N}$ where d_i represents the degree of a vertex v_i Table 7 lists the average degree of the three reply networks.

Table 7. Average degrees of the reply networks

Thread ID	Connected nodes		The whole network	
	Average degree	Size of authors	Average degree	Size of authors
1	2.818	2758	1.586	4890
2	2.445	2166	0.959	5514
3	1.958	4207	1.358	6065

4.3 Clustering Coefficient

Clustering coefficient is a measure of the degree to which nodes in a graph tend to cluster together[3]. The local clustering coefficient C_i for a vertex iis given by the proportion of links between the vertices within its neighborhoods divided by the number of links that could possibly exist between them [7]. The local clustering coefficient for directed graphs is given as $C_i = \frac{|\{e_{jk} : v_j, v_k \in N_i, e_{jk} \in E\}|}{k_i(k_i-1)}$, the edge e_{jk} connects vertex v_j and v_k. k_i is the number of neighbors of a vertex v_i. If $k_i = 0$ or $k_i = 1$, $C_i = 0$. The overall level of clustering in a network is measured as the average of the local clustering coefficients of all the vertices.

Table 8 lists the clustering coefficients of the three reply networks. The clustering coefficient of Network 2 is lower than others, which means the authors interact less in Network 2.

Table 8. Clustering coefficients of the reply networks

Thread ID	Authors interact with others		Reply networks	
	Clustering coefficient	Size of authors	Clustering coefficient	Size of authors
1	0.323	2758	0.182	4890
2	0.145	2166	0.057	5514
3	0.254	4207	0.176	6065

From the viewpoints of percentage of connected components, average indegree and clustering coefficient, the participants of Network 2 interact less frequently than participants of the Network 1 and Network 3. This is because that the original posts'

---

[2] https://en.wikipedia.org/wiki/Degree_(graph_theory).

[3] https://en.wikipedia.org/wiki/Clustering_coefficient.

stances about TCM are different. The original posts of Thread 1 and Thread 3 hold the same stance of "abolishing TCM". The expressions of the two original posts provoke heated arguments about TCM. The original post of Thread 2 holds the "preserving TCM" stance. The author of the original post claims that she is a daughter of a TCM practitioner, so some participants join in to ask for advices about therapies. It is inferred that these authors in Network 2 do not interact with others as frequently as authors in Network 1 and Network 3 do.

5   Triads Analysis

This section examines the characteristics of online reply networks at the triad level.

5.1   The Coded Triads

Davis and Leinhardt proposed that social relations could be tested on directed rather than undirected triad census, and forwarded 16 different types of directed triads as shown in Fig. 2 [8]. Their classification scheme describes each triad by a string of four elements: the number of mutual dyads within the triad; the number of asymmetric dyads within the triad; the number of null dyads within the triad; a configuration coding (U for up, D for down, C for cyclical and T for transitive) the triads which are not uniquely distinguished by the first three distinctions. For example, 120D refers the triad includes 1 mutual dyad, 2 asymmetric dyads, 0 null dyad and the down orientation.

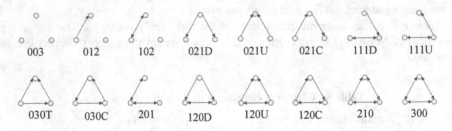

Fig. 2. 16 different types of directed triads [9]

We find the most active participants who post more than ten times in the three threaded discussions. Then we get three sub-graphs. Sub-graph 1 has 254 nodes and 1473 edges. Sub-graph 2 has 210 nodes and 1054 edges. Sub-graph 3 has 208 nodes and 854 edges. We use the "SNA" package[4] in R to calculate the types of the coded triads of these sub-graphs.

We add the proportions of 16 different types of directed triads respectively in the three sub-graphs. Table 10 lists the coded triads descending by their proportions.

---

[4] http://cran.r-project.org/web/packages/SNA/.

Table 9. Coded triads and their distributions

Coded triads	Sub-graph 1		Sub-graph 2		Sub-graph 3	
	Triads	‰	Triads	‰	Triads	‰
003	2323374	8608.26	1279216	8407.49	1295411	8763.10
012	158591	587.59	120101	789.35	91527	619.16
021C	5231	19.38	3219	21.16	2260	15.29
021D	4629	17.15	4810	31.61	2835	19.18
021U	4951	18.34	2281	14.99	3043	20.59
030C	27	0.10	24	0.16	23	0.16
030T	365	1.35	512	3.37	242	1.64
102	134097	496.84	83743	550.39	61637	416.96
111D	13079	48.46	7243	47.60	3407	23.05
111U	20503	75.97	7775	51.10	8836	59.77
120C	503	1.86	345	2.27	208	1.41
120D	367	1.36	698	4.59	142	0.96
120U	871	3.23	339	2.23	349	2.36
201	27173	100.68	8728	57.36	7174	48.53
210	2867	10.62	1505	9.89	743	5.03
300	2376	8.80	981	6.45	419	2.83

Table 10. Coded triads descending by their distributions

No.	Coded triads	‰	No.	Coded triads	‰
1	003	25778.85	9	021U	53.92
2	012	1996.09	10	210	25.54
3	102	1464.19	11	300	18.09
4	201	206.57	12	120U	7.82
5	111U	186.84	13	120D	6.91
6	111D	119.11	14	030T	6.35
7	021D	67.94	15	120C	5.54
8	021C	55.83	16	003C	0.42

Excluding the normal triads coded "003", "012", the triads coded "102", "201", "111U" or "111D" with mutual dyads outnumber other triads. The most common forms of the relationships between celebrities and audience or fans are one way relationships while the most common forms of the relationships between close friends are two way relationships [10]. In this paper, we can infer that the common forms in debate networks are mutual dyads.

5.2   Triadic Closure

Triadic closure process is one of the fundamental processes of structure formation. There is an increased chance that a friendship will form between two persons if they already have a friend in common [11].

To disclose the behavior patterns of participants with different opinions, we divide the participants in the debate into different groups by their stances: an "abolishing TCM" camp and a "preserving TCM" camp. We manually label the stances of authors by reading all their replies. We study the characteristics of the two main opposite groups in this sub-network. We find that nobody change their opinions during the debate. We compare the coded triadic closures in the two opposite camps. We choose the coded triads in triadic closure patterns: "030C", "030T", "120C", "120D", "210" and "300" in Fig. 2. Table 11 lists the size of the coded triadic closures and the ratio of these triads. From Table 11, we can see that a obviously higher proportion of directed triadic closures is in the "abolishing TCM" camp than that in the "preserving TCM" camp.

Table 11. Coded triadic closures in different camps

Thread ID	Preserving TCM		Abolishing TCM	
	Coded triadic closures	‰	Coded triadic closures	‰
1	141	5.66	67	183.28
2	123	9.48	69	97.91
3	100	5.00	58	70.33

5.3   Structural Balance

According to the balance theory, some social relations are more stable than others. For the triads, a friend of my friend is possibly more of my friend than my enemy. Balance is achieved when there are three positive links or two negatives with one positive. Two positive links and one negative creates imbalance as shown in Fig. 3.

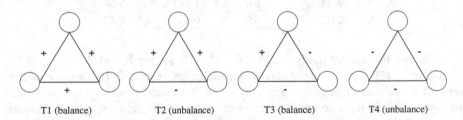

T1 (balance)   T2 (unbalance)   T3 (balance)   T4 (unbalance)

Fig. 3. Balanced and unbalanced triadic relationship [12]

We classify the edges into two types, the inter-camp edges whose end-points belong to different camps, and the inner-camp edges whose end-points come from the same camp. Then triads can be classified into two types "inner-camp" triads and "inter-camp" triads. We calculate the triads coded "300" (in Fig. 2) in the three

Table 12. Balanced "300" triads

Sub-graph	"preserving TCM" camp	"abolishing TCM" camp	inner-camp	inter-camp
1	48	312	360	2016
2	43	113	156	825
3	10	42	52	367

sub-graphs including the active participants in Table 9. Table 12 lists the sizes of the balanced "300" triads in the two camps and the size of the "inner-camp" balanced "300" triads (T1 in Fig. 3) and "inter-camp" balanced "300" triads (T3 in Fig. 3). Obviously, the balanced triads between camps are more than those within camps.

An interesting characteristic of many newsgroups is that people are more likely to respond to a message which they disagree than which they agree. This behavior is in sharp contrast to the WWW link graph, where linkage is an indicator of agreement or common interest [5]. This paper verifies the founding of Agrawal et al. from the perspective of structural balance.

6 Conclusions

This study explores the characteristics of online debate networks. Taking the debate on TCM at Tianya Forum as instance, we mine the interaction patterns at three levels.

At the individual level, some users reply to different threads related to the same topic. Authors of the original posts are the key players in reply networks. Some relationships among these authors are consistent across networks.

At the whole network level, the stance of an author who starts a thread affects the structural indices of the reply network. In our corpus, authors of original posts of Thread 1 and Thread 3 hold the "abolishing TCM" stance and the author of original post of Thread 2 holds the "preserving TCM" stance. That may explain why the percentage of connected nodes, average indegree, clustering coefficient of Network 2 are the lowest.

At the triad level, we find that mutual dyads are common form relationship in debate networks; balanced triads between the two camps are more than those within camps; there are more triadic closures within the "abolishing TCM" camp, which means the participants holding this opinion are more active to interact with each other.

In the future, we will do more studies to identify the behavior patterns of participants within debates. We will also combine the reply network analysis and text analysis to explore how opposing perspectives and arguments are put forward.

Acknowledgments. This work is supported by the National Natural Science Foundation of China (Nos. 61473284 and 71371107).

References

1. Somasundaran, S., Wiebe, J.: Recognizing stances in ideological on-line debates. In: NAACL HLT 2010 Workshop on Computational Approaches to Analysis and Generation of Emotion in Text, Los Angeles, pp. 116–124 (2010)
2. Liu, B.: Opinion mining and sentiment analysis. Found. Trends Inf. Retrieval 2(1–2), 1–135 (2008)
3. Wang, C., Tang, X.J.: Stance analysis for debates on traditional Chinese medicine at Tianya Forum. In: 5th International Conference on Computational Social Networks, Ho Chi Minh, pp. 321–332 (2016)
4. Agrawal, R., Rajagopalan, S., Srikant, R., et al.: Mining newsgroups using networks arising from social behavior. In: Proceedings of the 12th International Conference on World Wide Web, pp. 529–535. ACM, Budapest (2003)
5. Yang, Y., Chen, Q., Liu, W.: The structural evolution of an online discussion network. Phys. A Stat. Mech. Appl. 389(24), 5871–5877 (2010)
6. Zhao, Y.L., Tang, X.J.: In-depth analysis of online hot discussion about TCM. In: 15th International Symposium on Knowledge and Systems Science, pp. 275–283. JAIST Press, Sapporo (2014)
7. Watts, D.J., Strogatz, S.H.: Collective dynamics of 'small-world' networks. Nature 393 (6684), 440–442 (1998)
8. Davis, J.A., Leinhardt, S.: The structure of positive interpersonal relations in small groups. Sociological Theories in Progress 2, 218–251 (1972)
9. Li, Z.P., Tang, X.J.: Triadic and M-cycles balance in social network. In: 15th International Symposium on Knowledge and Systems Science, pp. 275–283. JAIST Press, Sapporo (2014)
10. Lou, T.C., Tang, J., Hopcroft, J., et al.: Learning to predict reciprocity and triadic closure in social networks. ACM Trans. Knowl. Discov. Data 7(2), 1825–1835 (2013)
11. Granovetter, M.S.: The strength of weak ties. Soc. Sci. Electron. Publishing 78(2), 1360–1380 (2015)
12. Li, Z.P., Tang, X.J.: Group polarization: connecting, influence and balance, a simulation study based on Hopfield modeling. In: 12th Pacific Rim International Conference on Artificial Intelligence, Kuching, pp. 710–721 (2012)

Entropy Measures for Extended Hesitant Fuzzy Linguistic Term Sets and Their Applications

Xia Liang[1], Cuiping Wei[2(✉)], and Xiaoyan Cheng[2]

[1] School of Management Science and Engineering,
Shandong University of Finance and Economics, Jinan 250014, China
[2] College of Mathematical Sciences, Yangzhou University, Yangzhou 225002, China
wei_cuiping@aliyun.com

Abstract. As a useful format of assessment, hesitant fuzzy linguistic term set (HFLTS) comes to the fore because of its wide applications in decision making. Extended hesitant fuzzy linguistic term set (EHFLTS) is a generalized form of HFLTS, and applied to collect the hesitant fuzzy linguistic information of a group. In this paper, we focus on the problem of how to measure the uncertain information of an EHFLTS. We first establish axiomatic definition of entropy for EHFLTS. Then we propose a series of entropy measures for EHFLTSs, which are also applicable for measure the uncertainty of HFLTSs. Compared with the entropy measures for HFLTSs provided by Farhadinia, the proposed entropy measures can measure both fuzziness and hesitation of an EHFLTS or HFLTS. Finally, we present a method, based on entropy and distance of EHFLTS, to determine the weights of the criteria for multi-criteria group decision making problem.

Keywords: Multi-criteria group decision making · Extended hesitant fuzzy linguistic term sets · Entropy · Weight

1 Introduction

In multi-criteria decision making (MCDM), due to the characteristics of qualitative criteria and limited attention of decision makers, decision makers usually feel more comfortable expressing imprecise information that represents their evaluations of the alternatives [1,2]. Generally, a linguistic term is one of the common formats for the expression of imprecise information [3–5].

In some practical MCDM problems, experts usually hesitate among different linguistic terms because of their uncertainty. Thus, it needs richer linguistic expressions to depict experts' assessments [6]. A recent proposal was introduced by Rodriguez et al. [7] to improve the elicitation of linguistic information in decision making by using hesitant fuzzy linguistic term sets (HFLTS) when experts hesitate among several linguistic terms to express their assessments. Following this research, HFLTSs have been investigated by researchers. And there several methods to solve the MCDM problem with HFLTSs information [8–10].

© Springer Nature Singapore Pte Ltd. 2016
J. Chen et al. (Eds.): KSS 2016, CCIS 660, pp. 151–164, 2016.
DOI: 10.1007/978-981-10-2857-1_13

The most common methods are based on hesitant fuzzy linguistic aggregated operators [11–13]. Moreover, there are some methods based on distance, correlation measure and similarity measure to solve MCDM problem with HFLTSs [14–16]. Besides, some traditional approaches, such as VIKOR method, TOPSIS method, outranking method and TODIM method are used for solving MCDM problem with hesitant fuzzy linguistic information, subsequently [17–20].

An HFLTS on a linguistic term set S is a subset with ordered consecutive linguistic terms in S. However, in multi-criteria group decision making (MCGDM) problems, we need use a nonconsecutive linguistic term set to describe the opinions of the multiple experts. This nonconsecutive linguistic term set is defined by Wang [21] as an extended hesitant fuzzy linguistic term set (EHFLTS). EHFLTSs can be used to collect the linguistic information of a group, and avoid the possible loss of information. More theory and methods need to be proposed. Wang [21] defined some basic operations for EHFLTSs. Then he developed two classes of aggregation operators for aggregating a set of EHFLTSs, so as to solve MCGDM problem with EHFLTs information. Wei et al. [22] defined some operations and distance measures of EHFLTSs. Then they developed a novel method to deal with extended hesitant fuzzy linguistic information.

Entropy is a very important notion for measuring uncertainty of fuzzy information. There are a lot of research productions of entropy measures for fuzzy sets, intuitionistic fuzzy sets and hesitant fuzzy sets [23–25]. This paper is focused on the problem that how to measure the uncertainty of an EHFLTS. ELTLTS is a simple and effective tool used to collect the linguistic information of a group. Suppose two experts compare the importance degree of two alternatives A and B on a linguistic term set $S = \{s_0 : \text{very unimportant}, s_1 : \text{unimportant}, s_2 : \text{a little unimportant}, s_3 : \text{equal}, s_4 : \text{a little important}, s_5 : \text{important}, s_6 : \text{very important}\}$, and one of them thinks that A is a little important than B, and the other thinks that A is very important than B. Then the importance degree of A compared to B might be described by the EHFLTS $\{s_4, s_6\}$, which describes the uncertain information of experts preference. Therefore, it is necessary to depict the uncertainty degree associated to the EHFLTS $\{s_4, s_6\}$ in an appropriate way. Similar to the conditional fuzzy set, the fuzziness is also an important index for EHFLTS. Besides, EHFLTS has its own characteristic, that is the hesitation among several linguistic term. Thus, in order to depict the uncertainty reflected by an EHFLTS, the fuzziness and hesitation of information should be considered. Therefore, the proposed axiomatic definitions and entropy measures should depict both fuzziness and hesitation of EHFLTS.

For HFLTSs, the special form of EHFLTSs, Farhadinia [26] presented some entropy measures based on distance measures and similarity measures. Although the research has made a significant contribution for HFLTS theory, it has some limitations. These entropies only measure the fuzziness degree of an HFLTS, that is the deviation of an HFLTS from the most fuzzy element. In fact, not only the fuzziness, frequently, the hesitation of the HFLTS should be also considered, which is illustrated in the following comparing Sect. 3.2 by examples. Moreover, these entropies is not appropriate for a general EHFLTS. Up to now, as far as we know, there has been few report concerning the entropy measure for EHFLTSs.

Motivated by these challenges, the purpose of this study is to investigate the entropy measure for EHFLTSs, and then to solve MCGDM problem with HFLTs information, in which the weights of criteria are completely unknown. First, considering the fuzziness and hesitation of EHFLTSs, an axiomatic definition of entropy and new entropy measures for EHFLTSs are developed. Further, based on entropy measures, a novel method is proposed to solve the MCGDM problem with HFLTs, in which the weights of criteria are completely unknown.

The remainder of this paper is structured as follows. Some basic preliminaries of EHFLTSs are introduced in Sect. 2. In Sect. 3, a novel axiomatic definition of entropy for EHFLTSs is proposed. Subsequently, a family of entropy measures for EHFLTSs is established. These proposed entropy measures are also appropriate to the special form HFLTSs, and so are compared with the existing ones for HFLTSs. In Sect. 4, based on the proposed entropy measures and the distance measures, a method for determining weights of criteria is proposed, so as to solve the MCGDM problem with HFLTSs. In Sect. 5, an example of selecting desirable global supplier(s) is presented to illustrate the feasibility and effectiveness of the proposed method. Conclusions are given in Sect. 6.

2 Preliminaries

2.1 Extended HFLTSs and Basic Operations

Herrera [27,28] introduced a finite and totally ordered linguistic term set $S = \{s_0, s_1, \cdots, s_g\}$ with odd cardinality and the mid term representing an assessment of "approximately 0.5". The rest of the terms are placed symmetrically around S. For example, a set S with seven terms could be given as: $S = \{s_0$: nothing, s_1 : very low, s_2 : low, s_3 : medium, s_4 : high, s_5 : very high, s_6 : perfect$\}$. Further, it is usually required that the linguistic term set S should satisfy the following additional characteristics.

(1) There is a negative operator: $Neg(s_i) = s_{g-i}$, where $g + 1$ is the cardinality of the term set;
(2) The set is ordered: $s_i \leq s_j \iff i \leq j$. Therefore, there exists a maximization operator: $max(s_i, s_j) = s_i$ if $s_j \leq s_i$, and a minimization operator: $min(s_i, s_j) = s_i$ if $s_i \leq s_j$.

Definition 1 [7]. Let $S = \{s_0, s_1, \cdots, s_g\}$ be a linguistic term set. An HFLTS H_S on S is a subset with ordered and consecutive linguistic terms in S.

Wang [21] generalized the definition of HFLTSs as follows.

Definition 2 [21]. Let $S = \{s_0, s_1, \cdots, s_g\}$ be a linguistic term set. An extended HFLTS (EHFLTS) H_S on S is a subset with ordered linguistic terms in S.

Suppose the elements in an EHFLTS are arranged in increasing order. Obviously, an HFLTS is a special EHFLTS in which the linguistic terms are ordered and consecutive [22].

For EHFLTSs, Wang [21] defined the "∨" and "∧" operations, which are called max-union and min-intersection operations in this paper, respectively.

Definition 3 [21]. Let $S = \{s_0, s_1, \cdots, s_g\}$ be a linguistic term set. For EHFLTSs H_S, H_S^1 and H_S^2 on S,

(1) $\{s_{g-i} \mid s_i \in H_S\}$ is the negation of H_S, denoted by $Neg(H_S)$;
(2) $\{max\{s_i, s_j\} \mid s_i \in H_S^1, s_j \in H_S^2\}$ is the max-union of H_S^1 and H_S^2, denoted by $H_S^1 \vee H_S^2$;
(3) $\{min\{s_i, s_j\} \mid s_i \in H_S^1, s_j \in H_S^2\}$ is the min-intersection of H_S^1 and H_S^2, denoted by $H_S^1 \wedge H_S^2$.

2.2   Distance Measures for EHFLTSs

The distance measure for EHFLTs is an effective tool to describe the deviation degree between two EHFLTs. Wei et al. [22] proposed the axiomatic definition of distance measure for EHFLTs, which is as following.

Definition 4 [22]. Let $S = \{s_0, s_1, \cdots, s_g\}$ be a linguistic term set, H_S^1 and H_S^2 be two EHFLTSs on S. Then the distance measure between H_S^1 and H_S^2, denoted by $d(H_S^1, H_S^2)$, should satisfy the following properties:

(1) $0 \leq d(H_S^1, H_S^2) \leq 1$;
(2) $d(H_S^1, H_S^2) = 0$ if and only if $H_S^1 = H_S^2$;
(3) $d(H_S^1, H_S^2) = d(H_S^2, H_S^1)$.

Based on the well-known Hamming distance, the Euclidean distance, the Hausdorff metric and Definition 4, Wei et al. [22] defined the following distance measures for EHFLTSs H_S^1 and H_S^2 with the same length l:
a normalized Hamming distance measure for EHFLTSs:

$$d_1(H_S^1, H_S^2) = \frac{1}{l} \sum_{k=1}^{l} \frac{|\Delta(s_{H_S^1}^k) - \Delta(s_{H_S^2}^k)|}{g}, \tag{1}$$

a normalized Euclidean distance measure for EHFLTSs:

$$d_2(H_S^1, H_S^2) = \left(\frac{1}{l} \sum_{k=1}^{l} \left(\frac{\Delta(s_{H_S^1}^k) - \Delta(s_{H_S^2}^k)}{g}\right)^2\right)^{\frac{1}{2}} \tag{2}$$

and a normalized Hausdorff distance measure for EHFLTS:

$$d_3(H_S^1, H_S^2) = \max_k \frac{|\Delta(s_{H_S^1}^k) - \Delta(s_{H_S^2}^k)|}{g}, \tag{3}$$

where $\Delta(s_i) = i$ and g is determined by the linguistic term set $S = \{s_0, s_1, \cdots, s_g\}$.

3   Entropy Measure for EHFLTSs and Comparison

In order to measure the uncertainty degree of an EHFLTS, it need to investigate entropy measure for an EHFLTS. First, a new axiomatic definition of the entropy measure for EHFLTSs is proposed. Furthermore, we put forward a series of entropy measures for EHFLTSs. Finally, the efficiency of the proposed entropy measures is demonstrated by comparing them with some existing entropy measures [26].

3.1   Entropy Measures for EHFLTs

For an EHFLTS $H_S = \{s^1_{H_S}, s^2_{H_S}, \cdots, s^l_{H_S}\}$ on the linguistic term sets $S = \{s_0, s_1, \cdots, s_g\}$, whose uncertainty of information includes two aspects: the fuzziness and hesitation of information. The fuzziness is dominated by the difference between the averaging value of H_S and the most fuzzy linguistic term $\{s_{g/2}\}$. The hesitation is reflected by the deviation degree of the elements in H_S. The averaging value and the deviation function value of H_S are defined as follows.

Definition 5. Let $H_S = \{s^1_{H_S}, s^2_{H_S}, \cdots, s^l_{H_S}\}$ be an EHFLTS. Then
(1) the averaging value (score function value) of H_S is defined as

$$\theta(H_S) = s_k, \tag{4}$$

where $k = \frac{1}{l}\sum_{i=1}^{l} \Delta(s^i_{H_S})$ and $\Delta(s^i_{H_S})$ is the subscript of $s^i_{H_S}$.
(2) the deviation function value of H_S is defined as

$$\eta(H_S) = \frac{2}{l(l-1)} \sum_{i=1}^{l-1} \sum_{j=i+1}^{l} (\Delta(s^j_{H_S}) - \Delta(s^i_{H_S})). \tag{5}$$

Based on the averaging value $\theta(H_S)$ and the deviation function value $\eta(H_S)$, an axiomatic definition of entropy for an EHFLTS is proposed.

Definition 6. Let $\mathbb{H}_S$ be the set of all the EHFLTSs on the linguistic term set S. The entropy on an EHFLTS H_S is a real-valued function $E : \mathbb{H}_S \to [0, 1]$, satisfying the following axiomatic requirements:

(E1) $E(H_S) = 0$, if and only if $H_S = \{s_0\}$ or $H_S = \{s_g\}$;
(E2) $E(H_S) = 1$, if and only if $\theta(H_S) = s_{g/2}$;
(E3) $E(H^1_S) \leq E(H^2_S)$, if $\theta(H^1_S) \leq \theta(H^2_S)$ and $\eta(H^1_S) \leq \eta(H^2_S)$ for $\theta(H^2_S) \leq s_{g/2}$,
    or $\theta(H^1_S) \geq \theta(H^2_S)$ and $\eta(H^1_S) \leq \eta(H^2_S)$ for $\theta(H^2_S) \geq s_{g/2}$;
(E4) $E(H_S) = E(H^c_S)$, where H^c_S is the complement of an EHFLT H_S and is defined by $H^c_S = \{s_{g-i}|s_i \in H_S\}$.

In the following, a family of entropy measures for EHFLTS based on the entropy measure E is presented. Additionally, three concrete entropy measures are studied.

Theorem 1. For each EHFLTS H_S, let

$$E(H_S) = \frac{f(\Delta(\theta(H_S))) + \alpha\eta(H_S)/g}{1 + \alpha\eta(H_S)/g}, \tag{6}$$

where $\alpha \in (0,1]$ is the contribution ratio of the averaging value and the deviation function value for uncertainty of H_S. Then $E(H_S)$ is an entropy measure for an EHFLTS H_S, where the function $f : [0,g] \to [0,1]$ satisfies the following three conditions:

1. $f(g-x) = f(x)$;
2. $f(x)$ is strictly monotone increasing with respect to $x \in [0,g/2]$ and strictly monotone decreasing with respect to $x \in [g/2,g]$;
3. it interpolates three points, $(0,0)$, $(g/2,1)$ and $(g,0)$.

Proof. It is sufficient to show the mapping $E(H_S)$, defined by Eq. (6), satisfying the conditions (E1)–(E4) in Definition 6.

For an EHFLTS $H_S = \{s^1_{H_S}, s^2_{H_S}, \cdots, s^l_{H_S}\}$, since $\theta(H_S) = s_k$, where $k = \frac{1}{l}\sum_{i=1}^{l}\Delta(s^i_{H_S})$ and $\eta(H_S) = \frac{2}{l(l-1)}\sum_{i=1}^{l-1}\sum_{j=i+1}^{l}(\Delta(s^j_{H_S}) - \Delta(s^i_{H_S}))$, we have $0 \le \Delta(\theta(H_S)) \le g$, and $0 \le \eta(H_S) \le g$. Since $0 \le f(x) \le 1$, we can obtain that

$$0 \le E(H_S) = \frac{f(\Delta(\theta(H_S))) + \alpha\eta(H_S)/g}{1 + \alpha\eta(H_S)/g} \le 1.$$

(E1) $E(H_S) = \frac{f(\Delta(\theta(H_S))) + \alpha\eta(H_S)/g}{1 + \alpha\eta(H_S)/g} = 0$ if and only if $f(\Delta(\theta(H_S))) = 0$ and $\eta(H_S) = 0$. According to the properties of the function $f(x)$, we can obtain that $E(H_S) = 0$ if and only if $H_S = \{s_0\}$ or $H_S = \{s_g\}$.

(E2) $E(H_S) = \frac{f(\Delta(\theta(H_S))) + \alpha\eta(H_S)/g}{1 + \alpha\eta(H_S)/g} = 1$ if and only if $f(\Delta(\theta(H_S))) = 1$. By the properties of function $f(x)$, we have $E(H_S) = 1$ if and only if $\theta(H_S) = s_{g/2}$.

(E3) Let $\Delta(\theta(H_S)) = \Delta$, $\eta(H_S) = \eta$, the Eq. (6) can be noted as $E(H_S) = E(\Delta, \eta) = \frac{f(\Delta) + \alpha\eta/g}{1 + \alpha\eta/g}$.

On one hand, the partial derivative of $E(\Delta, \eta)$ with respect to Δ is as

$$\frac{\partial E(\Delta, \eta)}{\partial \Delta} = \frac{f(\Delta)(1 + \alpha\eta/g)}{(1 + \alpha\eta/g)^2} = \frac{f'(\Delta)}{1 + \alpha\eta/g} \tag{7}$$

Since $\alpha \in (0,1]$ and $\eta \in (0,g]$, by the characteristics of $f(\Delta)$, we can see that $E(\Delta, \eta)$ is strictly monotone increasing with respect to $\Delta \in [0,g/2]$, and strictly monotone decreasing with respect to $\Delta \in [g/2,g]$.

On the other hand, the partial derivative of $E(\Delta, \eta)$ with respect to η is

$$\frac{\partial E(\Delta, \eta)}{\partial \eta} = \frac{\alpha/g(1 + \alpha\eta/g) - (f(\Delta) + \alpha\eta/g)\alpha/g}{(1 + \alpha\eta/g)^2} = \frac{\alpha(1 - f(\Delta))/g}{(1 + \alpha\eta/g)^2}.$$

Since $0 \le f(\Delta) \le 1$ and $\alpha \in (0,1]$, we have $\frac{\partial E(\Delta, \eta)}{\partial \eta} \ge 0$ and $E(\Delta, \eta)$ is monotone increasing with respect to $\eta \in [0,g]$.

Thus, it is easy to get that if $\theta(H_S^1) \leq \theta(H_S^2) \leq s_{g/2}$ and $\eta(H_S^1) \leq \eta(H_S^2)$, then $E(\theta(H_S^1),\ \eta(H_S^1)) \leq E(\theta(H_S^1),\ \eta(H_S^2)) \leq E(\theta(H_S^2),\ \eta(H_S^2))$, that is, $E(H_S^1) \leq E(H_S^2)$; if $\theta(H_S^1) \geq \theta(H_S^2) \geq s_{g/2}$ and $\eta(H_S^1) \leq \eta(H_S^2)$, then $E(\theta(H_S^1),\ \eta(H_S^1)) \leq E(\theta(H_S^1),\eta(H_S^2)) \leq E(\theta(H_S^2),\ \eta(H_S^2))$, i.e. $E(H_S^1) \leq E(H_S^2)$.

(E4) By Eq. (6), we have

$$E(H_S^c) = \frac{f(\Delta(\theta(H_S^c))) + \alpha\eta(H_S^c)/g}{1 + \alpha\eta(H_S^c)/g} = \frac{f(g - \Delta(\theta(H_S))) + \alpha\eta(H_S)/g}{1 + \alpha\eta(H_S)/g}$$

$$= \frac{f(\Delta(\theta(H_S))) + \alpha\eta(H_S)/g}{1 + \alpha\eta(H_S)/g} = E(H_S).$$

Equation (6) satisfies conditions (E1)–(E4) in Definition 6, thus E is an entropy measure for EHFLTS.

Corollary 1. $E(H_S)$ is strictly monotone increasing with respect to $\Delta(\theta(H_S)) \in [0, g/2]$ and strictly monotone decreasing with respect to $\Delta(\theta(H_S)) \in [g/2, g]$. In addition, it is monotone increasing with respect to $\eta(H_S) \in [0, g]$.

The conclusion is obviously shown in the proof process of Theorem 1.

It is noted that if we change the function $f(x)$ in $E(H_S)$ defined by Eq. (6), we can obtain a series of entropy measures for EHFLTs. For instance, let $\alpha = 1$ and let $f(x) = 1 - |1 - 2x/g|$, $f(x) = \sin(x/g \cdot \pi)$ and $f(x) = 1 - 4(x/g - \frac{1}{2})^2$, respectively. Three different entropy formulas are then obtained:

$$E_1(H_S) = \frac{1 - |1 - 2\Delta[\theta(H_S)]/g| + \eta(H_S)/g}{1 + \eta(H_S)/g}, \tag{8}$$

$$E_2(H_S) = \frac{\sin(\Delta[\theta(H_S)]/g \cdot \pi) + \eta(H_S)/g}{1 + \eta(H_S/g)}, \tag{9}$$

$$E_3(H_S) = \frac{1 - 4(\Delta[\theta(H_S)]/g - \frac{1}{2})^2 + \eta(H_S)/g}{1 + \eta(H_S)/g}. \tag{10}$$

3.2 Comparisons with Existing Entropy Measures

In this subsection, we compare the proposed entropy measures with the existing ones in order to illustrate the superiority of the entropy measures in our study. Farhadinia [26] proposed an axiomatic definition of HFLTSs on linguistic terms set $S' = \{s_{-\tau}, \cdots, s_{-1}, s_0, s_1, \cdots, s_\tau\}$. Considering the distance degree between an HFLTS and a fairness HFLTS s_0, an entropy measure is established, which denoted by $E_{F_1^\lambda}$. Besides, considering the similarity degree between an HFLTS and a complement HFLTS, an other new entropy measure is also proposed, which denoted by $E_{F_2^\lambda}$.

For the convenience of comparison, we present the transformation forms of the two entropy measures of HFLTS $H_S = \{s_{H_1}, s_{H_2}, \cdots, s_{H_L}\}$ on the linguistic terms set $S = \{s_0, s_1, \cdots, s_g\}$, which are as following.

$$E_{F_1^\lambda} = 1 - 2\left(\frac{1}{L}\sum_{l=1}^{L}\left(\frac{|\Delta(s_{H_l}) - g/2|}{g}\right)^\lambda\right)^{\frac{1}{\lambda}}; \tag{11}$$

$$E_{F_2^\lambda} = 1 - \left(\frac{1}{L} \sum_{l=1}^{L} \left(\frac{|\Delta(s_{H_l}) - g/2|}{g/2} \right)^\lambda \right)^{\frac{2}{\lambda}}, \qquad (12)$$

where $\Delta(s_{H_l})$ is the subscript of the linguistic term s_{H_l}.

Example 1. Let

$$H_S^1 = \{s_0\}, \ H_S^2 = \{s_1\}, \ H_S^3 = \{s_0, s_1, s_2\}, \ H_S^4 = \{s_2\},$$

$$H_S^5 = \{s_3\}, \ H_S^6 = \{s_2, s_3, s_4\}, \ H_S^7 = \{s_1, s_2, s_3, s_4, s_5\}, H_S^8 = \{s_4\}$$

be eight HFLTSs on the linguistic term set $S = \{s_0 : \text{nothing}, s_1 : \text{very low}, s_2 : \text{low}, s_3 : \text{slightly low}, s_4 : \text{medium}, s_5 : \text{slightly high}, s_6 : \text{high}, s_7 : \text{very high}, s_8 : \text{perfect}\}$.

The entropies of these HFLTSs are calculated by the entropy measures E_1, E_2, E_3, $E_{F_1^\lambda}$ and $E_{F_2^\lambda}$, respectively. The results are showed in Table 1.

Table 1. The entropies of $H_S^i (i = 1, 2, \cdots, 8)$ by different entropy measures

	H_S^1	H_S^2	H_S^3	H_S^4	H_S^5	H_S^6	H_S^7	H_S^8
$\theta(H_S)$	s_0	s_1	s_1	s_2	s_3	s_3	s_3	s_4
$\eta(H_S)$	0	0	1.3333	0	0	1.3333	2	0
$E_1(H_S)$	0	0.2500	0.3571	0.5000	0.7500	0.7857	0.8000	1
$E_2(H_S)$	0	0.3827	0.4709	0.7071	0.9239	0.9348	0.9391	1
$E_3(H_S)$	0	0.4375	0.5179	0.7500	0.9375	0.9464	0.9500	1
$E_{F_1^1}(H_S)$	0	**0.2500**	**0.2500**	0.3750	**0.7500**	**0.7500**	**0.7500**	1
$E_{F_2^1}(H_S)$	0	**0.4375**	**0.4375**	0.7500	**0.9375**	**0.9375**	**0.9375**	1

As can be seen from Table 1, the numerical examples in bold type reflect some counter-intuition cases with entropy measures $E_{F_1^\lambda}$ and $E_{F_2^\lambda}$. Take H_S^2 and H_S^3 for example, we can see that $\theta(H_S^2) = \theta(H_S^3)$, but $\eta(H_S^2) < \eta(H_S^3)$. Thus, H_S^3 is more uncertain than H_S^2. However, by Eqs. (11) and (12), we can see that $E_{F_1^1}(H_S^2) = E_{F_1^1}(H_S^3)$ and $E_{F_2^1}(H_S^2) = E_{F_2^1}(H_S^3)$, which is not consistent with our intuition. Also take H_S^5, H_S^6 and H_S^7 for example, the same limitations of $E_{F_1^\lambda}$ and $E_{F_2^\lambda}$ are demonstrated. Nevertheless, the entropy measures $E_i (i = 1, 2, 3)$ are effective to reflect hesitation and fuzziness of HFLTSs. Moreover, the proposed entropies $E_i (i = 1, 2, 3)$ can use to measure the uncertainty of a general EHFLTS, and $E_{F_1^\lambda}$ and $E_{F_2^\lambda}$ are not appropriate for EHFLTSs.

4   A Model Dealing with Hesitant Fuzzy Linguistic Group Decision Making Problems

Consider a multi-criteria linguistic group decision-making problem with hesitant fuzzy linguistic information. Let $A = \{A_1, A_2, \cdots, A_n\}$ with $n \geq 2$ be a finite set of alternatives and $C = \{C_1, C_2, \cdots, C_m\}$ with $m \geq 2$ be a finite set of criteria. Suppose there are t experts $D_1, D_2, \cdots, D_t$ to provide evaluations for alternatives A_i ($i = 1, 2, \cdots, n$) on criteria C_j ($j = 1, 2, \cdots, m$) using a linguistic term set $S = \{s_0, s_1, \cdots, s_g\}$. Suppose that the evaluation information of the kth expert is represented by a hesitant fuzzy linguistic decision matrix $R^k = (H_S^{ij(k)})_{n \times m}$, where $H_S^{ij(k)}$ is an HFLTS or a single linguistic term (which can be regarded as a special HFLTS) in S, and represents the linguistic assessment provided by the expert d_k for the alternative A_i on criterion C_j.

The problem addressed in the following study is how to assign the weights of criteria according to the hesitant fuzzy linguistic decision matrix $(H_S^{ij(k)})_{n \times m}$ and how to rank alternatives or to select the most desirable alternative(s). To elaborate, the decision process for alternatives is represented as follows.

4.1 Aggregate the Hesitant Fuzzy Linguistic Information of Experts

In the above linguistic group decision making problem, the HFLTS $H_S^{ij(k)}$ represents the evaluation information of the kth expert for the alternative A_i on the criterion C_j. We will use EHFLTSs to collect the evaluations of t experts based on the method provided by Wei et al. [22]. Let $H_S^{ij} = \bigcup_{k=1}^{t} H_S^{ij(k)}$, and for any two HFLTSs H_S^1 and H_S^2, $H_S^1 \bigcup H_S^2 = \{s_i | s_i \in H_S^1 \text{ or } H_S^2\}$. Then H_S^{ij} is an EHFLTS and represents the collective evaluation of t experts for the alternative A_i on the criterion C_j. From the decision matrices $R^k = (H_S^{ij(k)})_{n \times m}$, $k = 1, 2, \cdots, t$, we can construct a collective decision matrix $R = (H_S^{ij})_{n \times m}$.

Compared with existing method for aggregating collective decision matrix R, the method in our study use EHFLTSs to collect evaluations of all the experts, which has some advantages [22]. First, the method can eliminate the aggregation step on individual decision matrices. Moreover, the obtained EHFLTS decision matrix involves all the possible evaluations of all experts, and thus avoids the possible loss or distortion of information. In addition, experts can express their opinions by flexible forms such as single linguistic terms or HFLTSs, which is beneficial for experts to make evaluations in real applications.

4.2 Assign the Objective Weights of Criteria

In order to fuse the evaluations of all criteria for alternatives, we should determine the weights of criteria, which play a dominant role toward the final ranking of the alternatives. In general, the weights of criteria are predefined by the decision maker, which are regarded as one kind of subjective weights of the criteria. Sometimes, however, the subjective information about weights of criteria is completely unknown. Thus, it is a critical work to determine the weights of criteria according to assessment information, which are called objective weights.

Wei et al. [22] proposed a method, based on deviation degrees, to determine the objective weights of criteria. In the following, we improve the method of determining weights of criteria. Not only the deviation degrees, but also the uncertainty degrees of evaluations are considered to derive the objective weights of criteria in our study.

During the practical decision-making process, on one hand, the bigger the deviation degree under the criterion, the more important the criterion acts for the overall aggregation values. So we should assign a bigger weight to this criterion. On the other hand, we expect that the uncertainty degree of the assessment is as small as possible. Thus, the smaller the uncertainty degree under the criterion, the more important the criterion acts for the overall aggregation values.

Using the proposed entropy measures and distance measures, we now develop a method to determine the objective weights of criteria from the collective decision matrix $R = (H_S^{ij})_{n \times m}$. For alternative A_i on criterion C_j, the evaluation H_S^{ij} is an EHFLTS. Based on Eqs. (1), (2) or (3), we calculate the distance between H_S^{ij} and H_S^{lj}, which is denoted by $d(H_S^{ij}, H_S^{lj})$. Based on Eq. (6), we calculate the entropy of H_S^{ij}, which is denoted by E_{ij}. According to the above analysis, we give the following steps to determine the objective weights of criteria.

(1) Calculate the objective weights ω_j^1 of criteria C_j by

$$\omega_j^1 = \frac{\sum_{i=1}^{n-1} \sum_{l=i+1}^{n} d(H_S^{ij}, H_S^{lj})}{\sum_{j=1}^{m} \sum_{i=1}^{n-1} \sum_{l=i+1}^{n} d(H_S^{ij}, H_S^{lj})}, \quad j = 1, 2, \cdots, m. \tag{13}$$

(2) Calculate the objective weights ω_j^2 of criteria C_j by

$$\omega_j^2 = \frac{1 - \sum_{i=1}^{n} E_{ij}}{n - \sum_{j=1}^{m} \sum_{i=1}^{n} E_{ij}}, \quad j = 1, 2, \cdots, m. \tag{14}$$

(3) Integrate ω_j^1 and ω_j^2 into the weights ω_j of the criteria C_j

$$\omega_j = \gamma \omega_j^1 + (1 - \gamma) \omega_j^2, \quad \gamma \in [0, 1], \quad j = 1, 2, \cdots, m. \tag{15}$$

4.3 Rank Alternatives

In order to obtain the ranking result of alternatives, in the following, we utilize the extended hesitant fuzzy linguistic TOPSIS method proposed by Wei et al. [22] to derive the relative closeness degrees of alternatives, so as to obtain the ranking result. Originally proposed by Hwang and Yoon [29], the TOPSIS method is an effective tool for dealing with MCDM problems. TOPSIS simultaneously considers the distances from both the positive-ideal and the negative-ideal solutions. The alternatives are ranked by the closeness coefficients combining the two distances.

First, it is necessary to find the positive-ideal solution and negative-ideal solution. Let J_1 and J_2 be the subscripts of benefit criteria and cost criteria for the criteria set C, respectively. Suppose that $H^+ = \{\langle C_j, H_j^+ \rangle | C_j \in C\}$ is the hesitant fuzzy linguistic positive-ideal solution, and $H^- = \{\langle C_j, H_j^- \rangle | C_j \in C\}$ is the hesitant fuzzy linguistic negative-ideal solution. The elements H_j^+ and H_j^- ($j = 1, 2, \cdots, m$) are defined as follows:

$H_j^+ = H_S^{1j} \vee H_S^{2j} \vee \cdots \vee H_S^{nj}$ if C_j is a benefit criterion and $H_j^+ = H_S^{1j} \wedge$
$H_S^{2j} \wedge \cdots \wedge H_S^{nj}$ if C_j is a cost criterion;

$H_j^- = H_S^{1j} \wedge H_S^{2j} \wedge \cdots \wedge H_S^{nj}$ if C_j is a benefit criterion and $H_j^- = H_S^{1j} \vee$
$H_S^{2j} \vee \cdots \vee H_S^{nj}$ if C_j is a cost criterion.

For the alternative A_i, let $R_i = (H_S^{i1}, H_S^{i2}, \cdots, H_S^{im})$, $i = 1, 2, \cdots, n$. Then
calculate the weighted distances $d(R_i, H^+)$ and $d(R_i, H^-)$:

$$d(R_i, H^+) = \sum_{j=1}^m \omega_j d(H_S^{ij}, H_j^+), \qquad d(R_i, H^-) = \sum_{j=1}^m \omega_j d(H_S^{ij}, H_j^-). \qquad (16)$$

Further, calculate the closeness coefficients z_i of alternatives $A_i (i = 1, 2, \cdots, n)$:

$$z_i = \frac{d(R_i, H^+)}{d(R_i, H^-) + d(R_i, H^+)}. \qquad (17)$$

Finally, the alternatives are ordered according to the relative closeness
degrees. The smaller the closeness coefficients z_i is, the better the corresponding
alternatives A_i is.

From the above discussion about the group decision method, we develop
the following Algorithm to solve the above group decision-making problem with
hesitant fuzzy linguistic information. The following steps show how to apply the
multi-criteria decision making method.

Step 1. Calculate the criterion weights by Eqs. (13)–(15).
Step 2. Calculate the distances of each alternative to the positive-ideal solution
$d(R_i, H^+)$, and the negative-ideal solution $d(R_i, H^-)$ by Eq. (16).
Step 3. Calculate the closeness coefficient z_i for each alternative by Eq. (17).
Step 4. Rank the alternatives A_i according to the values of z_i in ascending
order, and the smaller the value of z_i is, the better the alternative A_i is.

5   An Illustrative Example

We use an example of selecting a global supplier, which is adapted from Wei
et al. [22], to illustrate the proposed method. A manufacturing company searches
the best global supplier for one of its most critical parts used in assembling
process among three suppliers $A_i(i = 1, 2, 3)$. The criteria are capacity of the
production (C_1), capacity of accuracy (C_2), suppliers credibility (C_3) and cost
performance of the product (C_4). Suppose there are three experts D_1, D_2 and D_3
to evaluate suppliers on these criteria by using the linguistic term set $S - \{s_0 :$
nothing, s_1 : very low, s_2 : low, s_3 : medium, s_4 : high, s_5 : very high, s_6 : perfect}.
The decision matrices $R_k = (H_S^{ij(k)})_{3 \times 4}$ ($k = 1, 2, 3$) are as follows:

$$R_1 = \begin{pmatrix} \{s_4, s_5\} & \{s_3\} & \{s_4\} & \{s_3\} \\ \{s_5\} & \{s_5\} & \{s_3\} & \{s_4\} \\ \{s_4\} & \{s_3\} & \{s_5\} & \{s_3\} \end{pmatrix}, \qquad R_2 = \begin{pmatrix} \{s_5, s_6\} & \{s_2\} & \{s_3, s_4\} & \{s_2\} \\ \{s_5\} & \{s_5\} & \{s_3\} & \{s_4\} \\ \{s_3\} & \{s_2, s_3\} & \{s_5\} & \{s_1\} \end{pmatrix},$$

$$R_3 = \begin{pmatrix} \{s_4\}\ \{s_2\}\ \{s_4\}\ \{s_2\} \\ \{s_5\}\ \{s_5\}\ \{s_3\}\ \{s_3\} \\ \{s_3\}\ \{s_4\}\ \{s_5\}\ \{s_1\} \end{pmatrix}.$$

Suppose the weights of the criteria $C_j (j = 1, 2, 3, 4)$ are completely unknown, then the decision making process is given to get the ranking result of alternatives.

First, the overall evaluations $H_S^{ij} (i = 1, 2, 3; j = 1, 2, 3, 4)$ of A_i on C_j are formed directly by the union of three experts' evaluations. For example, three experts' evaluations of alternative A_1 with respect to criterion C_1 are $\{s_4, s_5\}$, $\{s_5, s_6\}$ and $\{s_4\}$, respectively. Then the overall evaluation is formed by $\{s_4, s_5\} \bigcup \{s_5, s_6\} \bigcup \{s_4\}$, which is equal to the EHFLTS $\{s_4, s_5, s_6\}$. So the collective decision matrix R is as follows:

$$R = \begin{pmatrix} \{s_4, s_5, s_6\}\ \{s_2, s_3\} & \{s_3, s_4\}\ \{s_2, s_3\} \\ \{s_5\} & \{s_5\} & \{s_3\}\ \{s_3, s_4\} \\ \{s_3, s_4\} & \{s_2, s_3, s_4\}\ \{s_5\} & \{s_1, s_3\} \end{pmatrix}.$$

Then, we calculate the objective weights of criteria. On one hand, suppose we adopt the distance measure d_2 defined by Eq. (2) and the decision maker is optimistic. Based on Eq. (13), we can obtain the weights of criteria as follows.

$$\omega^1 = (0.2286, 0.3170, 0.2572, 0.1972)^T. \tag{18}$$

On the other hand, suppose we adopt the entropy measure E_1 defined by Eq. (8), as well as Eq. (14). Then we can obtain the weights of criteria as follows.

$$\omega^2 = (0.3862, 0.2306, 0.2306, 0.1526)^T. \tag{19}$$

By Eq. (15), let $\gamma = 0.5$, we obtain the objective weighting vector of criteria

$$\omega = (0.3074, 0.2738, 0.2439, 0.1749)^T. \tag{20}$$

Further, we determine the positive-ideal and the negative-ideal solutions $H^+ = (\{s_5, s_6\}, \{s_5\}, \{s_5\}, \{s_3, s_4\})$, $H^- = (\{s_3, s_4\}, \{s_2, s_3\}, \{s_3\}, \{s_1, s_2, s_3\})$. By Eqs. (2) and (16), and suppose the decision maker is optimistic, we have

$$d_2(R_1, H^+) = 0.2512, \ d_2(R_2, H^+) = 0.1172, \ d_2(R_3, H^+) = 0.2462,$$

$$d_2(R_1, H^-) = 0.1244, \ d_2(R_2, H^-) = 0.2471, \ d_2(R_3, H^-) = 0.0981.$$

So the closeness coefficients z_i of the three suppliers are: $z_1 = 0.6689, z_2 = 0.3217, z_3 = 0.7150$, and the ranking is $A_2 \succ A_1 \succ A_3$. Thus, the desirable supplier is A_2.

6   Conclusions

Entropy is an important tool to measure uncertainty of fuzzy information and has broad application backgrounds. This paper presents an axiomatic definition of entropy for EHFLTSs, and constructs a class of entropy formulas. Based on the proposed entropy measures for EHFLTSs, we develop a method for solving MCGDM problem with hesitant fuzzy linguistic information. Compared with existing works, the studies in this paper has the following contributions. First, the notion of entropy for EHFLTSs is defined and some concrete entropy formulas are proposed to depict the uncertainty of the EHFLTSs. These entropy formulas are also appropriate for HFLTS. Besides, the proposed entropy formulas consider both the fuzziness and hesitancy of EHFLTSs and HFLTSs, and overcome the limitations found in existing entropy measures that only considering the fuzziness of HFLTSs. Second, based on the entropy measures of EHFLTSs, a method of determining the objective weights of criteria is proposed. Not only the deviation degrees, but also the uncertainty degrees of evaluations are considered to derived the objective weights of criteria in our study. It can be used in the situation that the weights of criteria are completely known. For future research, the entropy theory of EHFLTSs with probability distributions will be investigated.

Acknowledgements. The authors are most grateful to the referees and the editors for their constructive suggestions. The work was partly supported by the National Natural Science Foundation of China (71371107, 11271224), and the Top-notch Academic Programs Project of Jiangsu Higher Education Institutions (No. PPZY2015B109).

References

1. Zadeh, L.A.: Fuzzy sets. Inf. Control **8**, 338–356 (1965)
2. Herrera, F., Herrera-Viedma, E., Verdegay, J.L.: A model of consensus in group decision making under linguistic assessments. Fuzzy Sets Syst. **78**, 73–87 (1996)
3. Zadeh, L.A.: The concept of a linguistic variable and its applications to approximate reasoning. Inf. Sci. Part I **8**, 199–249; Part II **8**, 301–357; Part III **9**, 43–80 (1975)
4. Herrera, F., Martínez, L.: A model based on linguistic 2-tuples for dealing with multigranularity hierarchical linguistic contexts in multiexpert decision-making. IEEE Trans. Syst. Man Cybern. Part B Cybern. **31**, 227–234 (2001)
5. Rodríguez, R.M., Martínez, L.: An analysis of symbolic linguistic computing models in decision making. Int. J. Gen. Syst. **42**, 121–136 (2013)
6. Herrera, F., Martínez, L.: A 2-tuple fuzzy linguistic representation model for computing with words. IEEE Trans. Fuzzy Syst. **8**, 746–752 (2000)
7. Rodriguez, R.M., Martínez, L., Herrera, F.: Hesitant fuzzy linguistic term sets for decision making. IEEE Trans. Fuzzy Syst. **20**, 109–119 (2012)
8. Zhu, B., Xu, Z.S.: Consistency measures for hesitant fuzzy linguistic preference relations. IEEE Trans. Fuzzy Syst. **22**, 35–45 (2014)
9. Montes, R., Snchez, A.M., Villar, P., Herrera, F.: A web tool to support decision making in the housing market using hesitant fuzzy linguistic term sets. Appl. Soft Comput. **35**, 949–957 (2015)
10. Wei, C.P., Zhao, N., Tang, X.J.: Operations and comparisons of hesitant fuzzy linguistic term sest. IEEE Trans. Fuzzy Syst. **22**, 575–585 (2014)
11. Wu, J.T., Wang, J.Q., Wang, J., Zhang, H.Y., Chen, X.H.: Hesitant fuzzy linguistic multicriteria decision-making method based on generalized prioritized aggregation operator. Sci. World J. **2014**, 1–16 (2014)

12. Yang, S.H., Ju, Y.B.: Dual hesitant fuzzy linguistic aggregation operators and their applications to multi-attribute decision making. J. Intell. Fuzzy Syst. **27**, 1935–1947 (2014)

13. Lin, R., Zhao, X.F., Wang, H.J., Wei, G.W.: Hesitant fuzzy linguistic aggregation operators and their application to multiple attribute decision making. J. Intell. Fuzzy Syst. **36**, 49–63 (2014)

14. Hesamian, G., Shams, M.: Measuring similarity and ordering based on hesitant fuzzy linguistic term sets. J. Intell. Fuzzy Syst. **28**, 983–990 (2015)

15. Liao, H.C., Xu, Z.S.: Approaches to manage hesitant fuzzy linguistic information based on the cosine distance and similarity measures for HFLTSs and their application in qualitative decision making. Expert Syst. Appl. **42**, 5328–5336 (2015)

16. Xu, Y.J., Xu, A.W., Merig, J.M., Wang, H.M.: Hesitant fuzzy linguistic ordered weighted distance operators for group decision making. J. Appl. Math. Comput. **49**, 1–24 (2015)

17. Liao, H.C., Xu, Z.S., Zeng, X.J.: Hesitant fuzzy linguistic VIKOR method and its application in qualitative multiple criteria decision making. IEEE Trans. Fuzzy Syst. **23**, 1343–1355 (2015)

18. Beg, I., Rashid, T.: TOPSIS for hesitant fuzzy linguistic term sets. Int. J. Intell. Syst. **28**, 1162–1171 (2013)

19. Wang, J., Wang, J.Q., Zhang, H.Y., Chen, X.H.: Multi-criteria decision-making based on hesitant fuzzy linguistic term sets: an outranking approach. Knowl. Based Syst. **86**, 224–236 (2015)

20. Wei, C.P., Ren, Z.L., Rodrguez, R.M.: A hesitant fuzzy linguistic TODIM method based on a score function. Int. J. Comput. Intell. Syst. **8**, 701–712 (2015)

21. Wang, H.: Extended hesitant fuzzy linguistic term sets and their aggregation in group decision making. Int. J. Comput. Intell. Syst. **8**, 14–33 (2014)

22. Wei, C.P., Zhao, N., Tang, X.J.: A novel linguistic group decision-making model based on extended hesitant fuzzy linguistic term sets. Int. J. Uncertainty Fuzziness Knowl. Based Syst. **23**, 379–398 (2015)

23. De, L.A., Termini, S.: A definition of nonprobabilistic entropy in the setting of fuzzy sets theory. Inf. Control **20**, 301–312 (1972)

24. Liang, X., Wei, C.P., Xia, M.M.: New entropy, similarity measure of intuitionistic fuzzy sets and their applications in group decision making. Int. J. Comput. Intell. Syst. **6**, 987–1001 (2013)

25. Wei, C.P., Yan, F.F., Rodrígueza, R.M.: Entropy measures for hesitant fuzzy sets and their application in multi-criteria decision-making. J. Intell. Fuzzy Syst. **31**, 673–685 (2016)

26. Farhadinia, B.: Multiple criteria decision-making methods with completely unknown weights in hesitant fuzzy linguistic term setting. Knowl. Based Syst. **93**, 135–144 (2015)

27. Herrera, F., Herrera-Viedma, E., Verdegay, L.: A sequential selection process in group decision making with linguistic assessment approach. Inf. Sci. **85**, 223–239 (1995)

28. Herrera, F., Herrera-Viedma, E.: A model of consensus in group decision making under linguistic assessments. Fuzzy Sets Syst. **78**, 73–87 (1996)

29. Hwang, C.L., Yoon, K.: Multiple Attributes Decision Making Methods and Applications. Springer, Heidelberg (1981)

Group Decision Making Based on Acceptably Consistent Interval Multiplicative Preference Relations

Zhen Zhang[1](✉), Wenyu Yu[2], and Chonghui Guo[1]

[1] Institute of Systems Engineering, Dalian University of Technology,
Dalian 116024, People's Republic of China
{zhen.zhang,dlutguo}@dlut.edu.cn
[2] Institute of Information Management and Information Systems,
Dalian University of Technology, Dalian 116024, People's Republic of China
ywyeva@mail.dlut.edu.cn

Abstract. In this paper, group decision making problems based on interval multiplicative preference relations are investigated. First, the acceptable consistency property for an interval multiplicative preference relation is discussed, based on which some optimization models are established to derive an acceptably consistent interval multiplicative preference relation based on the initial one. The interval multiplicative preference relation derived by the proposed model has the minimum deviation from the initial one and can preserve the decision information as much as possible. Subsequently, an approach to group decision making with interval multiplicative preference relations is developed based on the proposed models. Eventually, a numerical example is provided to illustrate the proposed approach.

Keywords: Group decision making · Interval multiplicative preference relation · Consistency · Optimization

1 Introduction

Group decision making is more and more common in human being's daily life due to the increasing complexity of the socio-economic environment. For group decision making problems, the preference relation is an effective tool for decision makers to elicit their preference information over alternatives, because it only needs decision makers to compare two alternatives at one time [8]. In the last few decades, different types of preference relations have been proposed and used to deal with group decision making problems [1,7,8,13,16].

The multiplicative preference relation, initially proposed by Saaty [8] and further used in his famous AHP [9], is one of the most widely used types of preference relations. To handle the uncertainty of the decision environment and the impreciseness of decision makers' judgments, the interval multiplicative preference relation (IMPR) is usually used in practical decision making problems [10].

© Springer Nature Singapore Pte Ltd. 2016
J. Chen et al. (Eds.): KSS 2016, CCIS 660, pp. 165–174, 2016.
DOI: 10.1007/978-981-10-2857-1_14

In general, the research on IMPRs can be mainly grouped into four categories: (1) deriving priority weight vectors from an individual IMPR or a group of IMPRs [2,14]; (2) checking and improving the consistency for an individual IMPR [3,4]; (3) managing incomplete information in decision making with IMPRs [6,12]; (4) consensus building models for group decision making problems with IMPRs [19].

Among the four categories, consistency is an important issue due to the fact that inconsistent IMPRs can result in unreasonable and misleading decision results [4,15,17]. Although different proposals have been devoted to address the consistency issue for an IMPR [3,4,6,11], it is still an open question which needs to be further studied. The main focus of this paper is to develop some new models to derive an acceptably consistent IMPR from the initial one and then apply the proposed models to deal with group decision making problems.

The remainder of this paper is structured as follows. In Sect. 2, some preliminaries are provided. Afterwards, some optimization models are developed to derive an acceptably consistent IMPR from the initial one in Sect. 3. Section 4 proposes an approach to group decision making based on the models. To demonstrate the proposed approach, a numerical example is provided in Sect. 5. Finally, this paper is concluded in Sect. 6.

2 Preliminaries

In this section, some preliminaries about IMPRs are revised. For convenience, let $I = \{1, 2, \ldots, n\}$.

Saaty [8] defined the multiplicative preference relation (also called a pairwise comparison matrix) as follows.

Definition 1. [8] $A = (a_{ij})_{n \times n}$ *is called a multiplicative preference relation if* $a_{ij} > 0$, $a_{ii} = 1$, $a_{ij}a_{ji} = 1$, $i, j \in I$.

Different numerical scales have been developed for multiplicative preference relations, and the most widely used scale is the 1–9 scale given by Saaty [8], i.e., $a_{ij} \in [1/9, 9]$, which will be used throughout this paper.

Definition 2. [9] *A multiplicative preference relation* $A = (a_{ij})_{n \times n}$ *is considered completely consistent, if* $a_{ik}a_{kj} = a_{ij}$ *for all* $i, j, k \in I$.

To handle the uncertainty and impreciseness existing in practical decision making problems, the IMPR was defined by Saaty and Vargas [10].

Definition 3. $\tilde{U} = (\tilde{u}_{ij})_{n \times n}$ *is called an interval multiplicative preference relation (IMPR) if* $\tilde{u}_{ij} = [u_{ij}^-, u_{ij}^+]$, $0 < u_{ij}^- \leq u_{ij}^+$, $u_{ii} = [1, 1]$, $u_{ij}^- u_{ji}^+ = 1$ *and* $u_{ji}^- u_{ij}^+ = 1$ *for all* $i, j \in I$.

Wang et al. [11] defined the acceptable consistency for an IMPR as follows.

Definition 4. *For an IMPR* $\tilde{U} = (\tilde{u}_{ij})_{n \times n}$, *where* $\tilde{u}_{ij} = [u_{ij}^-, u_{ij}^+]$, *if there exists a multiplicative preference relation* $A = (a_{ij})_{n \times n}$ *such that* $u_{ij}^- \leq a_{ij} \leq u_{ij}^+$ *and* $a_{ik}a_{kj} = a_{ij}$ *for all* $i, j, k \in I$, *then* $\tilde{U}$ *is acceptably consistent.*

3 Deriving Acceptably Consistent IMPRs

In this section, some optimization models are developed to derive an acceptably consistent IMPR.

Based on Wang et al.'s definition [11], Dong and Herrera-Videma [5] provided the following proposition to check whether an IMPR is acceptably consistent.

Proposition 1. *An IMPR* $\tilde{U} = (\tilde{u}_{ij})_{n \times n}$, *where* $\tilde{u}_{ij} = [u_{ij}^-, u_{ij}^+]$, *is of acceptable consistency if and only if for* $i, j, k \in I$

$$u_{ik}^+ u_{kj}^+ \geq u_{ij}^- \tag{1}$$

and

$$u_{ik}^- u_{kj}^- \leq u_{ij}^+. \tag{2}$$

In what follows, a simplified version of Proposition 1 is given.

Proposition 2. *An IMPR* $\tilde{U} = (\tilde{u}_{ij})_{n \times n}$, *where* $\tilde{u}_{ij} = [u_{ij}^-, u_{ij}^+]$, *is of acceptable consistency if and only if Eqs. (1) and (2) hold for all* $i < k < j$, *and* $i, j, k \in I$.

Proof. Necessity. By Proposition 1, if $\tilde{U}$ is acceptably consistent, Eqs. (1) and (2) hold for all $i, j, k \in I$. Hence, Eqs. (1) and (2) hold for $i < k < j$ and $i, j, k \in I$.

Sufficiency. If two or three elements of $\{i, j, k\}$ are equal, Eqs. (1) and (2) are reduced to the conditions of Definition 3. Therefore, it is only needed to prove the cases where i, j, k take different values.

Case 1: $i < k < j$. In this case, Eqs. (1) and (2) hold obviously.

Case 2: $i < j < k$. In this case, we have $u_{ij}^+ u_{jk}^+ \geq u_{ik}^-$ and $u_{ij}^- u_{jk}^- \leq u_{ik}^+$. As $u_{jk}^- u_{kj}^+ = 1$ and $u_{jk}^+ u_{kj}^- = 1$, it follows that $u_{ij}^+ \cdot \dfrac{1}{u_{kj}^-} \geq u_{ik}^-$ and $u_{ij}^- \cdot \dfrac{1}{u_{kj}^+} \leq u_{ik}^+$, i.e., $u_{ik}^- u_{kj}^- \leq u_{ij}^+$ and $u_{ik}^+ u_{kj}^+ \geq u_{ij}^-$.

Case 3: $k < i < j$. In this case, we have $u_{ki}^+ u_{ij}^+ \geq u_{kj}^-$ and $u_{ki}^- u_{ij}^- \leq u_{kj}^+$. As $u_{ki}^- u_{ik}^+ = 1$ and $u_{ki}^+ u_{ik}^- = 1$, it follows that $\dfrac{1}{u_{ik}^-} \cdot u_{ij}^+ \geq u_{kj}^-$ and $\dfrac{1}{u_{ik}^+} \cdot u_{ij}^- \leq u_{kj}^+$, i.e., $u_{ik}^- u_{kj}^- \leq u_{ij}^+$ and $u_{ik}^+ u_{kj}^+ \geq u_{ij}^-$.

Case 4: $k < j < i$. In this case, we have $u_{kj}^+ u_{ji}^+ \geq u_{ki}^-$ and $u_{kj}^- u_{ji}^- \leq u_{ki}^+$. As $u_{ik}^- u_{kj}^+ = 1$, $u_{ik}^+ u_{kj}^- = 1$, $u_{ij}^- u_{ji}^+ = 1$ and $u_{ij}^+ u_{ji}^- = 1$, it follows that $u_{kj}^+ \cdot \dfrac{1}{u_{ij}^-} \geq \dfrac{1}{u_{ik}^+}$ and $u_{kj}^- \cdot \dfrac{1}{u_{ij}^+} \leq \dfrac{1}{u_{ik}^-}$, i.e., $u_{ik}^+ u_{kj}^+ \geq u_{ij}^-$ and $u_{ik}^- u_{kj}^- \leq u_{ij}^+$.

Case 5: $j < i < k$. In this case, we have $u_{ji}^+ u_{ik}^+ \geq u_{jk}^-$ and $u_{ji}^- u_{ik}^- \leq u_{jk}^+$. As $u_{ji}^- u_{ij}^+ = 1$, $u_{ji}^+ u_{ij}^- = 1$, $u_{jk}^- u_{kj}^+ = 1$ and $u_{jk}^+ u_{kj}^- = 1$, it follows that $\dfrac{1}{u_{ij}^-} \cdot u_{ik}^+ \geq \dfrac{1}{u_{kj}^+}$ and $\dfrac{1}{u_{ij}^+} \cdot u_{ik}^- \leq \dfrac{1}{u_{kj}^-}$, i.e., $u_{ik}^+ u_{kj}^+ \geq u_{ij}^-$ and $u_{ik}^- u_{kj}^- \leq u_{ij}^+$.

Case 6: $j < k < i$. In this case, we have $u_{jk}^+ u_{ki}^+ \geq u_{ji}^-$ and $u_{jk}^- u_{ki}^- \leq u_{ji}^+$. As $u_{jk}^- u_{kj}^+ = 1$, $u_{jk}^+ u_{kj}^- = 1$, $u_{ki}^- u_{ik}^+ = 1$, $u_{ki}^+ u_{ik}^- = 1$, $u_{ji}^- u_{ij}^+ = 1$ and $u_{ji}^+ u_{ij}^- = 1$, it follows that $\dfrac{1}{u_{kj}^-} \cdot \dfrac{1}{u_{ik}^-} \geq \dfrac{1}{u_{ij}^+}$ and $\dfrac{1}{u_{ij}^+} \cdot \dfrac{1}{u_{ik}^+} \leq \dfrac{1}{u_{ij}^-}$, i.e., $u_{ik}^- u_{kj}^- \leq u_{ij}^+$ and $u_{ik}^+ u_{kj}^+ \geq u_{ij}^-$.

In summary, Eqs. (1) and (2) hold for all $i, j, k \in N$. As a result, $\tilde{U} = (\tilde{u}_{ij})_{n \times n}$ is acceptably consistent according to Proposition 1. This completes the proof of Proposition 2.

However, the IMPR provided by a decision maker is usually not of acceptable consistency. As a result, it is proposed some optimization models to derive an acceptably consistent IMPR based on the initial one.

Assume that there is an acceptably consistent IMPR $\tilde{U}^* = (\tilde{u}_{ij}^*)_{n \times n}$, where $\tilde{u}_{ij}^* = [u_{ij}^{*-}, u_{ij}^{*+}]$ and $u_{ij}^{*-} \leq u_{ij}^{*+}$ $i, j \in I$, then it follows that for all $i < k < j$ and $i, j, k \in I$,

$$u_{ik}^{*+} u_{kj}^{*+} \geq u_{ij}^{*-} \tag{3}$$

and

$$u_{ik}^{*-} u_{kj}^{*-} \leq u_{ij}^{*+}, \tag{4}$$

which can be rewritten as

$$\log_9 u_{ik}^{*+} + \log_9 u_{kj}^{*+} \geq \log_9 u_{ij}^{*-} \tag{5}$$

and

$$\log_9 u_{ik}^{*-} + \log_9 u_{kj}^{*-} \leq \log_9 u_{ij}^{*+}. \tag{6}$$

It is hoped that the deviation between $\tilde{U}$ and $\tilde{U}^*$ should be minimized. Based on this idea, the following optimization model is developed

$$\min \quad J = \sum_{i=1}^{n-1} \sum_{j=i+1}^{n} \left(\left| \log_9 u_{ij}^- - \log_9 u_{ij}^{*-} \right| + \left| \log_9 u_{ij}^+ - \log_9 u_{ij}^{*+} \right| \right)$$

$$\text{s.t.} \quad \log_9 u_{ik}^{*+} + \log_9 u_{kj}^{*+} \geq \log_9 u_{ij}^{*-}, i < k < j, \text{and } i, j, k \in I \qquad \text{(M-1)}$$

$$\log_9 u_{ik}^{*-} + \log_9 u_{kj}^{*-} \leq \log_9 u_{ij}^{*+}, i < k < j, \text{and } i, j, k \in I$$

$$\log_9(1/9) \leq \log_9 u_{ij}^{*-} \leq \log_9 u_{ij}^{*+} \leq \log_9 9, i < j, \text{and } i, j \in I.$$

The objective function of the model (M-1) is the logarithm Manhattan distance between $\tilde{U}$ and $\tilde{U}^*$. The first two constraints guarantee that $\tilde{U}^*$ is acceptably consistent.

To solve the model (M-1), the following proposition is provided.

Proposition 3. *By using some transformation variables g_{ij} and h_{ij}, the model (M-1) can be equivalently transformed into*

$$\min \quad J = \sum_{i=1}^{n-1} \sum_{j=i+1}^{n} (g_{ij} + h_{ij})$$

$$\text{s.t.} \quad \log_9 u_{ik}^{*+} + \log_9 u_{kj}^{*+} \geq \log_9 u_{ij}^{*-}, i < k < j, \text{and } i, j, k \in I$$

$$\log_9 u_{ik}^{*-} + \log_9 u_{kj}^{*-} \leq \log_9 u_{ij}^{*+}, i < k < j, and\ i,j,k \in I$$

$$g_{ij} \geq \log_9 u_{ij}^- - \log_9 u_{ij}^{*-}, g_{ij} \geq \log_9 u_{ij}^{*-} - \log_9 u_{ij}^-, i < j, and\ i,j \in I$$

$$h_{ij} \geq \log_9 u_{ij}^+ - \log_9 u_{ij}^{*+}, h_{ij} \geq \log_9 u_{ij}^{*+} - \log_9 u_{ij}^+, i < j, and\ i,j \in I$$

$$-1 \leq \log_9 u_{ij}^{*-} \leq \log_9 u_{ij}^{*+} \leq 1, i < j, and\ i,j \in I. \tag{M-2}$$

Proof. In the model (M-2), Lines 3–4 of the constraints guarantee that $g_{ij} \geq |\log u_{ij}^- - \log u_{ij}^{*-}|$ and $h_{ij} \geq |\log u_{ij}^+ - \log u_{ij}^{*+}|$, respectively. As the objective is to minimize the sum of all the g_{ij} and h_{ij}, any feasible solution which satisfies $g_{ij} > |\log u_{ij}^- - \log u_{ij}^{*-}|$ and $h_{ij} > |\log u_{ij}^+ - \log u_{ij}^{*+}|$ is not the optimal solution to the model (M-2). Consequently, $g_{ij} = |\log u_{ij}^- - \log u_{ij}^{*-}|$ and $h_{ij} = |\log u_{ij}^+ - \log u_{ij}^{*+}|$, $i < j, i, j \in I$, which yields $J = \sum_{i=1}^{n-1} \sum_{j=i+1}^{n} (g_{ij} + h_{ij}) = \sum_{i=1}^{n-1} \sum_{j=i+1}^{n} (|\log_9 u_{ij}^- - \log_9 u_{ij}^{*-}| + |\log_9 u_{ij}^+ - \log_9 u_{ij}^{*+}|)$.

Line 5 guarantees that $\log_9(1/9) \leq \log_9 u_{ij}^{*-} \leq \log_9 u_{ij}^{*+} \leq \log_9 9$, $i < j$, and $i, j \in I$.

By combining Lines 1–2, the model (M-1) can be equivalently described as the model (M-2). This completes the proof of Proposition 3.

The model (M-2) is a linear programming model whose decision variables are $\log_9 u_{ij}^{*-}$ and $\log_9 u_{ij}^{*+}$ and can be solved by using some optimization software packages. Let p_{ij} and q_{ij} be the optimal values of $\log_9 u_{ij}^{*-}$ and $\log_9 u_{ij}^{*+}$ derived by the model (M-2), then we have

$$u_{ij}^{*-} = 9^{p_{ij}}, u_{ij}^{*+} = 9^{q_{ij}}, u_{ji}^{*-} = 1/u_{ij}^{*+}, u_{ji}^{*+} = 1/u_{ij}^{*-}, i < j\ and\ i,j \in I. \tag{7}$$

By Definition 3, an acceptably consistent IMPR is derived. Moreover, the following proposition is easily obtained.

Proposition 4. *Let J^* be the optimal objective function value of the model (M-2), then an IMPR $\tilde{U} = (\tilde{u}_{ij})_{n \times n}$ is acceptably consistent if and only if $J^* = 0$.*

Remark 1. The value of J^* is the minimum adjustment amount that is needed to improve the consistency of an unacceptably consistent IMPR. In particular, if $J^* = 0$, the IMPR does not need to be adjusted. Therefore, Proposition 4 provides a new way to check the acceptable consistency for an IMPR $\tilde{U}$.

Sometimes, there may be more than one solution to the model (M-2). To address this issue, it is developed the following model to refine the solutions to the model (M-2):

$$\min \quad J' = \sum_{i=1}^{n-1} \sum_{j=i+1}^{n} (\log_9 u_{ij}^{*+} - \log_9 u_{ij}^{*-})$$

$$\text{s.t.} \quad \sum_{i=1}^{n-1} \sum_{j=i+1}^{n} (g_{ij} + h_{ij}) = J^*$$

$$\log_9 u_{ik}^{*+} + \log_9 u_{kj}^{*+} \geq \log_9 u_{ij}^{*-}, i < k < j, \text{and } i, j, k \in I \qquad \text{(M-3)}$$

$$\log_9 u_{ik}^{*-} + \log_9 u_{kj}^{*-} \leq \log_9 u_{ij}^{*+}, i < k < j, \text{and } i, j, k \in I$$

$$g_{ij} \geq \log_9 u_{ij}^{-} - \log_9 u_{ij}^{*-}, g_{ij} \geq \log_9 u_{ij}^{*-} - \log_9 u_{ij}^{-}, i < j, \text{and } i, j \in I$$

$$h_{ij} \geq \log_9 u_{ij}^{+} - \log_9 u_{ij}^{*+}, h_{ij} \geq \log_9 u_{ij}^{*+} - \log_9 u_{ij}^{+}, i < j, \text{and } i, j \in I$$

$$-1 \leq \log_9 u_{ij}^{*-} \leq \log_9 u_{ij}^{*+} \leq 1, i < j, \text{and } i, j \in I,$$

where J^* is the optimal objective function value of the model (M-2).

It is obvious that the model (M-3) aims to minimize the overall uncertainty of the IMPR $\tilde{U}^*$ under the constraint that the objective function value is equal to J^* and the constraints of the model (M-2).

Remark 2. Compared with previous studies about how to improve the consistency of an IMPR, the proposed models have the following advantages: (1) The proposed models are established based on the acceptable consistency of an IMPR, which do not need to determine a consistency threshold in advance; (2) The proposed models obtain an acceptably consistent IMPR by minimizing the overall adjustment amount of an IMPR, which not only can guarantee the reasonability of the decision result, but also can preserve the decision information as much as possible.

4   Group Decision Making Based on IMPRs

In this section, an approach to group decision making based on IMPRs is developed. For the convenience of analysis, the group decision making problem is formulated as below.

Let $X = \{x_1, x_2, \ldots, x_n\}$ be the set of alternatives, and $E = \{e^1, e^2, \ldots, e^m\}$ be the set of decision makers. To select the best alternative or rank the alternatives, each decision maker e^l gives his/her preference information as an IMPR $\tilde{U}^l = ([u_{ij}^{l-}, u_{ij}^{l+}])_{n\times n}, l \in L = \{1, 2, \ldots, m\}$. The weight vector of the decision makers is $\lambda = (\lambda_1, \lambda_2, \ldots, \lambda_m)^{\mathrm{T}}$, where $0 \leq \lambda_l \leq 1, l \in L, \sum_{l=1}^{m} \lambda_l = 1$.

Before proposing the approach to group decision making based on IMPRs, the consistency of the group IMPR is investigated.

Proposition 5. *Let $\tilde{U}^l$, $l \in L$ and λ be defined as before and aggregate all the individual IMPRs into a group IMPR $\tilde{U} = ([u_{ij}^{-}, u_{ij}^{+}])_{n\times n}$, where*

$$u_{ij}^{-} = \prod_{k=1}^{m}(u_{ij}^{l-})^{\lambda_l}, u_{ij}^{+} = \prod_{k=1}^{m}(u_{ij}^{l+})^{\lambda_l}, i, j \in I. \qquad (8)$$

If each $\tilde{U}^l$ is acceptably consistent, then $\tilde{U}$ is also acceptably consistent.

Proof. If each $\tilde{U}^l$ is acceptably consistent, then we have

$$u_{ik}^{l+}u_{kj}^{l+} \geq u_{ij}^{l-}, i < k < j, \text{and } i, j, k \in I$$

and

$$u_{ik}^{l-} u_{kj}^{l-} \leq u_{ij}^{l+}, i < k < j, \text{and } i, j, k \in I.$$

It follows that

$$(u_{ik}^{l+} u_{kj}^{l+})^{\lambda_l} \geq (u_{ij}^{l-})^{\lambda_l}, i < k < j, \text{and } i, j, k \in I \tag{9}$$

and

$$(u_{ik}^{l-} u_{kj}^{l-})^{\lambda_l} \leq (u_{ij}^{l+})^{\lambda_l}, i < k < j, \text{and } i, j, k \in I. \tag{10}$$

Multipling both sides of Eqs. (9) and (10) for all $l \in L$ yields

$$\prod_{l=1}^{m} (u_{ik}^{l+} u_{kj}^{l+})^{\lambda_l} \geq \prod_{l=1}^{m} (u_{ij}^{l-})^{\lambda_l}, \prod_{l=1}^{m} (u_{ik}^{l-} u_{kj}^{l-})^{\lambda_l} \leq \prod_{l=1}^{m} (u_{ij}^{l+})^{\lambda_l}, i < k < j, \text{and } i, j, k \in I,$$

which can be written as

$$\prod_{l=1}^{m} (u_{ik}^{l+})^{\lambda_l} \prod_{l=1}^{m} (u_{kj}^{l+})^{\lambda_l} \geq \prod_{l=1}^{m} (u_{ij}^{l-})^{\lambda_l}, i < k < j, \text{and } i, j, k \in I,$$

$$\prod_{l=1}^{m} (u_{ik}^{l-})^{\lambda_l} \prod_{l=1}^{m} (u_{kj}^{l-})^{\lambda_l} \leq \prod_{l=1}^{m} (u_{ij}^{l+})^{\lambda_l}, i < k < j, \text{and } i, j, k \in I,$$

i.e.,

$$u_{ik}^{+} u_{kj}^{+} \geq u_{ij}^{-}, u_{ik}^{-} u_{kj}^{-} \leq u_{ij}^{+}, i < k < j, \text{and } i, j, k \in I.$$

As a result, $\tilde{U}$ is also acceptably consistent according to Proposition 2. This completes the proof of Proposition 5.

Proposition 5 demonstrates that if all the individual IMPRs are acceptably consistent, then their fusion is also acceptably consistent. Therefore, it is reasonable to derive the ranking of alternatives from the group IMPR. Based on this idea, an approach to group decision making based on IMPRs is developed.

Step 1: Solve the model (M-2) for each individual IMPR $\tilde{U}^l$, $l \in L$ to derive the optimal objective function value for the IMPR $\tilde{U}^l$ as J^{l*}. If $J^{l*} = 0$ for all $l \in L$, go to Step 3; otherwise, go to Step 2.

Step 2: For the IMPRs whose J^{l*} is unequal to 0, employ the model (M-3) to derive an acceptably consistent IMPR. For convenience, the derived acceptably consistent IMPR is also denoted by $\tilde{U}^l$. Go to Step 3.

Step 3: Fuse all the individual IMPRs into a group one $\tilde{U}$ by Eq. (8).

Step 4: Calculate the interval-valued priority weight vector $w = (w_1, w_2, \ldots, w_n)^{\mathrm{T}}$ for the alternatives, where $w_i = [w_i^-, w_i^+]$, $i \in I$ such that

$$w_i^- = \prod_{j=1}^{m} \sqrt[n]{(u_{ij}^-)}, w_i^+ = \prod_{j=1}^{m} \sqrt[n]{u_{ij}^+}, i \in I. \tag{11}$$

Step 5: Rank the alternatives and select the best one.

Remark 3. In the proposed approach, the model (M-2) is solved to check the acceptable consistency for each IMPR $\tilde{U}^l$ based on Proposition 4 in Step 1. For each IMPR that is unacceptably consistent, the model (M-3) is further employed to derive an acceptably consistent IMPR in Step 2. Finally, all the individual IMPRs are aggregated into a group one, based on which the priority weight vector is calculated to rank the alternatives.

5 A Numerical Example

In this section, a numerical example is used to demonstrate the proposed group decision making approach.

Assume that three decision makers e^1, e^2 and e^3, whose weight vector is $\lambda = (0.3, 0.4, 0.3)^{\mathrm{T}}$, provide their preference information over a set of four alternatives $\{x_1, x_2, x_3, x_4\}$ using IMPRs as below [19]:

$$
\tilde{U}^1 = \begin{pmatrix}
[1,1] & [1,3] & [3,5] & [4,6] \\
[1/3,1] & [1,1] & [3,4] & [1/2,1] \\
[1/5,1/3] & [1/4,1/3] & [1,1] & [2,4] \\
[1/6,1/4] & [1,2] & [1/4,1/2] & [1,1]
\end{pmatrix}
$$

$$
\tilde{U}^2 = \begin{pmatrix}
[1,1] & [2,4] & [4,6] & [4,8] \\
[1/4,1/2] & [1,1] & [4,7] & [1/5,1/3] \\
[1/6,1/4] & [1/7,1/4] & [1,1] & [1,3] \\
[1/8,1/4] & [3,5] & [1/3,1] & [1,1]
\end{pmatrix},
$$

$$
\tilde{U}^3 = \begin{pmatrix}
[1,1] & [2,3] & [1,2] & [3,5] \\
[1/3,1/2] & [1,1] & [3,5] & [1,2] \\
[1/2,1] & [1/5,1/3] & [1,1] & [6,8] \\
[1/5,1/3] & [1/2,1] & [1/8,1/6] & [1,1]
\end{pmatrix}.
$$

First, the consistency of each IMPR is checked. By solving the model (M-2), it is found that the optimal objective function values for the three IMPRs are 0.8155, 1.1309, 1, respectively. Hence, all the three IMPRs are not acceptably consistent according to Proposition 4.

By solving the model (M-3) for the three IMPRs, it can be derived the corresponding acceptably consistent IMPRs as

$$
\tilde{U}^1 = \begin{pmatrix}
[1,1] & [1,3] & [3,5] & [4,6] \\
[1/3,1] & [1,1] & [1.2599,4] & [1/2,1.5874] \\
[1/5,1/3] & [1/4,0.7937] & [1,1] & [1.2599,4] \\
[1/6,1/4] & [0.6300,2] & [1/4,0.7937] & [1,1]
\end{pmatrix}
$$

$$
\tilde{U}^2 = \begin{pmatrix}
[1,1] & [2,4] & [4,6] & [4,8] \\
[1/4,1/2] & [1,1] & [1.400,7] & [1/5,1] \\
[1/6,1/4] & [1/7,0.7143] & [1,1] & [0.7143,3] \\
[1/8,1/4] & [1,5] & [1/3,1.400] & [1,1]
\end{pmatrix},
$$

$$\tilde{U}^3 = \begin{pmatrix} [1,1] & [2,3] & [1,2] & [3,5] \\ [1/3,1/2] & [1,1] & [1,5] & [1,5] \\ [1/2,1] & [1/5,1] & [1,1] & [5,8] \\ [1/5,1/3] & [1/5,1] & [1/8,1/5] & [1,1] \end{pmatrix}.$$

It can be checked that all the IMPRs derived by the model (M-3) are acceptably consistent. By fusing the three acceptably consistent IMPRs using Eq. (8), the group IMPR is derived as

$$\tilde{U} = \begin{pmatrix} [1,1] & [1.6245,3.3659] & [2.4208,4.0856] & [3.6693,6.3734] \\ [0.2971,0.6156] & [1,1] & [1.2262,5.3499] & [0.4267,1.8616] \\ [0.2448,0.4131] & [0.1869,0.8155] & [1,1] & [1.5182,4.3893] \\ [0.1569,0.2725] & [0.5372,2.3437] & [0.2278,0.6587] & [1,1] \end{pmatrix}.$$

By solving the model (M-2) for $\tilde{U}$, it is found that $\tilde{U}$ is also of acceptable consistency, which verifies Proposition 5.

The interval-valued priority weights for the alternatives are calculated by Eq. (11) as $w_1 = [1.9490, 3.0597]$, $w_2 = [0.6279, 1.5736]$, $w_3 = [0.5134, 1.1027]$, $w_4 = [0.3722, 0.8054]$.

According to the ranking method proposed in [6], the ranking of the alternatives is $x_1 \succ x_2 \succ x_3 \succ x_4$. Therefore, the best alternative is x_1.

6   Conclusions

The IMPR is an effective tool for decision makers to elicit their imprecise preference information for practical decision making problems. In this paper, it is firstly developed some optimization models to derive an acceptably consistent IMPR from the initial one, based on which an approach to group decision making with IMPRs is then proposed. The numerical example demonstrates the feasibility and effectiveness of the proposed approach.

In terms of future research, it is interesting to extend the proposed models and approach to handle group decision making problems with incomplete IMPRs or other types of preference relations, such as the hesitant fuzzy preference relation [18] and the hesitant fuzzy linguistic preference relation [20].

Acknowledgements. This work was supported by the National Natural Science Foundation of China (Nos. 71501023, 71171030), the Funds for Creative Research Groups of China (No. 71421001), the China Postdoctoral Science Foundation (2015M570248) and the Fundamental Research Funds for the Central Universities (DUT15RC(3)003).

References

1. Alonso, S., Cabrerizo, F.J., Chiclana, F., Herrera, F., Herrera-Viedma, E.: Group decision making with incomplete fuzzy linguistic preference relations. Int. J. Intell. Syst. **24**, 201–222 (2009). doi:10.1002/int.20332

2. Arbel, A.: Approximate articulation of preference and priority derivation. Eur. J. Oper. Res. **43**, 317–326 (1989). doi:10.1016/0377-2217(89)90231-2
3. Conde, E., de la Pérez, M.P.R.: A linear optimization problem to derive relative weights using an interval judgement matrix. Eur. J. Oper. Res. **201**, 537–544 (2010). doi:10.1016/j.ejor.2009.03.029
4. Dong, Y., Chen, X., Li, C.C., Hong, W.C., Xu, Y.: Consistency issues of interval pairwise comparison matrices. Soft Comput. **19**, 2321–2335 (2015). doi:10.1007/s00500-014-1426-2
5. Dong, Y., Herrera-Viedma, E.: Consistency-driven automatic methodology to set interval numerical scales of 2-tuple linguistic term sets and its use in the linguistic GDM with preference relation. IEEE Trans. Cybern. **45**, 780–792 (2015). doi:10.1109/TCYB.2014.2336808
6. Liu, F., Zhang, W.G., Wang, Z.X.: A goal programming model for incomplete interval multiplicative preference relations and its application in group decision-making. Eur. J. Oper. Res. **218**, 747–754 (2012). doi:10.1016/j.ejor.2015.06.015
7. Orlovsky, S.A.: Decision-making with a fuzzy preference relation. Fuzzy Set Syst. **1**, 155–167 (1978). doi:10.1016/0165-0114(78)90001-5
8. Saaty, T.L.: A scaling method for priorities in hierarchical structures. J. Math. Psychol. **15**, 234–281 (1977). doi:10.1016/0022-2496(77)90033-5
9. Saaty, T.L.: The Analytic Hierarchy Process. McGraw-Hill, New York (1980)
10. Saaty, T.L., Vargas, L.G.: Uncertainty and rank order in the analytic hierarchy process. Eur. J. Oper. Res. **32**, 107–117 (1987). doi:10.1016/0377-2217(87)90275-X
11. Wang, Y.M., Yang, J.B., Xu, D.L.: Interval weight generation approaches based on consistency test and interval comparison matrices. Appl. Math. Comput. **167**, 252–273 (2005). doi:10.1016/j.amc.2004.06.080
12. Wang, Z.J.: A note on "a goal programming model for incomplete interval multiplicative preference relations and its application in group decision-making". Eur. J. Oper. Res. **247**, 867–871 (2015). doi:10.1016/j.ejor.2015.06.015
13. Xu, Z.: Intuitionistic preference relations and their application in group decision making. Inf. Sci. **177**, 2363–2379 (2007). doi:10.1016/j.ins.2006.12.019
14. Xu, Z., Cai, X.: Deriving weights from interval multiplicative preference relations in group decision making. Group Decis. Negot. **23**, 695–713 (2014). doi:10.1007/s10726-012-9315-5
15. Xu, Z., Wei, C.: A consistency improving method in the analytic hierarchy process. Eur. J. Oper. Res. **116**, 443–449 (1999). doi:10.1016/S0377-2217(98)00109-X
16. Zhang, Z., Guo, C.: An approach to group decision making with heterogeneous incomplete uncertain preference relations. Comput. Ind. Eng. **71**, 27–36 (2014). doi:10.1016/j.cie.2014.02.004
17. Zhang, Z., Guo, C.: Consistency and consensus models for group decision-making with uncertain 2-tuple linguistic preference relations. Int. J. Syst. Sci. **47**, 2572–2587 (2016). doi:10.1080/00207721.2014.999732
18. Zhang, Z., Guo, C.: Fusion of heterogeneous incomplete hesitant preference relations in group decision making. Int. J. Comput. Intell. Syst. **9**, 245–262 (2016). doi:10.1080/18756891.2016.1149999
19. Zhang, Z., Guo, C.: Minimum adjustment-based consistency and consensus models for group decision making with interval pairwise comparison matrices. In: Proceedings of the 2016 IEEE International Conference on Fuzzy Systems (FUZZ-IEEE 2016), pp. 1701–1708. IEEE Press (2016)
20. Zhu, B., Xu, Z.: Consistency measures for hesitant fuzzy linguistic preference relations. IEEE T. Fuzzy Syst. **22**, 35–45 (2014). doi:10.1109/TFUZZ.2013.2245136

Implementation of the Distributed Fixed-Point Algorithm and Its Application

Zhengtian Wu[1]($\boxtimes$), Qinfen Shi[1], Yu Yu[2], Haili Xia[1], and Hongyan Yang[1]

[1] Suzhou University of Science and Technology, Suzhou, China
wzht8@mail.usts.edu.cn
[2] Nanjing Audit University, Nanjing, China

Abstract. A implementation of the distributed Dang and Ye's fixed-point algorithm, which is a new alternative algorithm for integer programming, is developed in this paper. This fixed-point algorithm is derived from an increasing mapping which satisfies certain properties. A classical problem, which is called market split problem, has been solved by this distributed implementation. It is shown that this implementation is effective in numerical results. Besides, it can be used to other similar integer problems.

Keywords: Fixed-point algorithm · Increasing mapping · Integer programming · Market split problem

1 Introduction

Integer programming, which is an NP-complete problem [1], has been playing a significant role in management and economics. Many contributions have been made in the following literatures. The cutting plane method, which iteratively improves a feasible set or objective function using linear inequalities, is originated in [2]. A branch-and-bound approach [3] is made up of a systematic enumeration of all the solutions which are candidate, where large fruitless candidate subsets are discarded at the same time, by means of upper and lower estimated bounds of the quantity which is being optimized. The neighborhood algorithm is developed for integer programming in [4]. Several new developments for integer programming can be found in the recent literatures such as [5,7]. These algorithms provide the basic research foundation in integer programming. However, it is still a challenge to solve integer problems effectively in a short time.

In this research, the implementation of the Dang and Ye's fixed-point algorithm as described in Fig. 1 is developed. The details of the implementation will be discussed and some numerical results can be obtained. A classic operational research problem, market split problem, has been tested by this distributed implementation. Some comparisons will be made between this distributed implementation method and other traditional methods. The numerical results and the comparisons show that this distributed implementation is effective.

© Springer Nature Singapore Pte Ltd. 2016
J. Chen et al. (Eds.): KSS 2016, CCIS 660, pp. 175–181, 2016.
DOI: 10.1007/978-981-10-2857-1_15

The organization of this paper is as follows. Some details of the implementation of this fixed-point algorithm will be introduced in Sect. 2. In Sect. 3, the market split problem will be solved by this distributed implementation. The details of the computing and numerical results can be obtained in this section. Finally, Sect. 4 will present conclusions of this paper.

2   Details of Implementation of the Fixed-point Method

In order to solve the integer problem effectively, a fixed-point method has been developed by Dang and Ye in [6,7]. The idea of this method can be explained as follows. Let $P = \{x \in R^n | Ax \leq b\}$, where $A \in R^{m \times n}$ is an $m \times n$ integer matrix with $n \geq 2$, and b a vector of R^m. Let $x^{max} = (x_1^{max}, x_2^{max}, \ldots, x_n^{max})^T$ with $x_j^{max} = max_{x \in P} x_j, j = 1, 2, \ldots, n$ and $x^{min} = (x_1^{min}, x_2^{min}, \ldots, x_n^{min})^T$ with $x_j^{min} = min_{x \in P} x_j, j = 1, 2, \ldots, n$. Let $D(P) = \{x \in Z^n | x^l \leq x \leq x^u\}$, where $x^l = \lfloor x^{min} \rfloor$ and $x^u = \lfloor x^{max} \rfloor$. For $z \in R^n$ and $k \in N_0$, let $P(z, k) = \{x \in P | x_i = z_i, 1 \leq i \leq k, and\ x_i \leq z_i, k + 1 \leq i \leq n\}$. Given an integer point $y \in D(P)$ with $y_1 > x_i^l$. the fixed-point algorithm can be described as in Fig. 1.

In order to solve integer problem, Dang and Ye developed a fixed-point method in which a increasing-mapping is defined from a finite lattice into itself. In the lexicographical order, all the integer points which are outside the polytope P are mapped to the first point in polytope P. Under this increasing mapping, every integer point that is inside the polytope is the fixed point. In this method, within limited number of iterations this procedure either proves there is no such fixed-point exists in polytope P or yields an integer point after given any initial integer point. For more information about this algorithm, one can see [6,7].

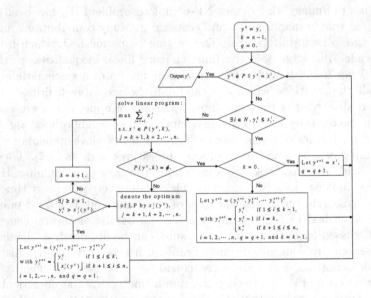

Fig. 1. Flow diagram of the iterative method.

As an appeal feature, Dang and Ye's fixed-point algorithm could be easily run in a distributed mode. Some details of distributed implementation to Dang and Ye's algorithm will be introduced in this section.

This distributed computing system consists of one master computer and several number of slave computers. The tasks of the master computer can be listed as follows.

(1) Solving the solution space of the polytope.

(2) Segmenting the solution space into several small subpolytope according to the number of slave computers.

(3) Sending each subpolytope to one slave computer.

(4) Receiving the slave computers' computation result.

(5) Sorting out the computation result and outputting the result.

And the tasks of each slave computer can be listed as follows.

(1) Receiving the subpolytope.

(2) Solving the fixed point in its subpolytope by using Dang and Ye's fixed-point algorithm in Fig. 1.

(3) Sending each computation result to the master computer.

The process of the distributed computation can be described in the Fig. 2.

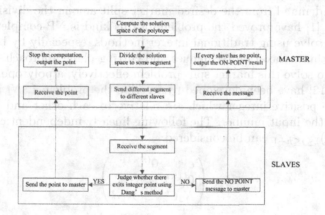

Fig. 2. Flow diagram of the distributed computation process.

In this distributed implementation, the simplex algorithm [8] has been used to the bounding linear programming in each iterative process. A freely available software, which is called MPICH2, is brought in to receive and send message between the slaves and the master. The detail information of Message Passing Interface is given in the literature [9].

3   Numerical Results

Here is the typical example of the market split problem:

Example 1. A company consists of two divisions which are called D_1 and D_2. The retailers are supplied with several products from this company. Each retailer is allocated to either division D_1 or division D_2. The goal is to make sure that division D_1 owns the 40 percent of the company's market for every product, and division D_2 owns the remaining or, for all the products if such a 40/60 split is not possible, to minimize the sum of percentage deviations from the 40/60 split.

A set of 0-1 linear programming has been formulated to describe the market split problem in [10]. The mathematics expression about the feasibility version of the problem could be described in (1).

$$\exists \quad x_j \in \{0,1\}, \quad j = 1, 2, \ldots, n$$
$$s.t. \quad \sum_{j=1}^{n} c_{ij}x_j = d_i, i = 1, 2, \ldots, m? \tag{1}$$

where m denotes the products' number, n denotes the retailers' number, c_{ij} denotes the retailer j's demand for product i, and vector d_i, which is in the right hand side, is demand from the desired market split among the division D_1 and divisions D_2. [1] have proved the problem of this kind is NP-complete and it is very hard to solve using traditional integer methods, especially by branch-and-bound algorithm, relaxation techniques and cutting plane approach.

In order to solve this market split problem effectively, a polytope judgement problem [11,12] have been developed to describe the problem 1. N_1 and N_2 are defined to be positive integer which are relative to each other and big enough according to the input number. The following linearly independent column vectors $\boldsymbol{B} = (\boldsymbol{b}_j)_{1 \leq j \leq n+1}$ can be considered:

$$\boldsymbol{B} = \begin{pmatrix} e^{(n)} & \mathbf{0}^{(n \times 1)} \\ \mathbf{0}^{(1 \times n)} & N_1 \\ N_2 \boldsymbol{C} & -N_2 \boldsymbol{d} \end{pmatrix}$$

where $\boldsymbol{C}$ represents $(c_{ij})_{m \times n}$ in (1), $\boldsymbol{d}$ represents $(d_i)_m$ in (1), $e^{(n)}$ represents the n-dimensional identity matrix and $\mathbf{0}^{n \times 1}$ denotes the $n \times 1$-dimension zero matrix. The LLL basis reduction approach that is popular polynomial approach is used to the lattice $\boldsymbol{B}$, then one can be obtained the reformulation as follows:

$$\hat{\boldsymbol{B}} = \begin{pmatrix} \boldsymbol{X}_0^{(n \times (n-m))} & \boldsymbol{x}_d \\ \mathbf{0}^{(1 \times (n-m))} & N_1 \\ \mathbf{0}^{(m \times (n-m))} & \mathbf{0}^{(m \times 1)} \end{pmatrix}$$

Lemma 1. *(Aardal et al. [11]) The problem (1) can be reformulated equivalently as:*
Is there exist a vector

$$\boldsymbol{\lambda} \in Z^{n-m} \quad s.t. - \boldsymbol{x}_d \leq \boldsymbol{X}_0 \boldsymbol{\lambda} \leq e^{(n-m) \times 1} - \boldsymbol{x}_d \tag{2}$$

Table 1. Four methods to solve market split problem

| P. | Solving system 2 after basis reduction | | | | | | Solving system 1 directly | | | | | |
| | The method | | BC | | BB | | BC | | BB | | CP | |
	NumLPs	F	NumLPs	F	NumLPs	F	NumLPs	F	NumLPs	F	NumLPs	F
1	9670	Feasible	4334	Feasible	1.50E+06	Feasible	5.00E+07	Feasible	*	*	E+09	Feasible
2	32015	Feasible	36624	Feasible	1.2 E+06	Feasible	6.70E+07	Feasible	*	*	E+09	Feasible
3	22221	Feasible	23652	Feasible	9.2 E+05	Feasible	7.90E+07	Feasible	*	*	E+09	Feasible
4	12670	Feasible	6924	Feasible	1.2 E+06	Feasible	5.30E+07	Feasible	*	*	2.59E+07	Feasible
5	49709	Feasible	22163	Feasible	2.5 E+06	Feasible	2.60E+06	Feasible	*	*	1.24E+09	Feasible
6	54525	Infeasible	34501	Infeasible	3.4 E+06	Infeasible	2.60E+08	Infeasible	*	*	E+09	Infeasible
7	105570	Infeasible	86555	Infeasible	4.60E+06	Infeasible	1.60E+08	Infeasible	*	*	E+09	Infeasible
8	90204	Infeasible	77661	Infeasible	5.00E+06	Infeasible	8.20E+07	Infeasible	*	*	E+09	Infeasible
9	93751	Infeasible	53586	Infeasible	5.10E+06	Infeasible	2.20E+08	Infeasible	*	*	E+09	Infeasible
10	67565	Infeasible	51837	Infeasible	3.3 E+06	Infeasible	1.10E+08	Infeasible	*	*	E+09	Infeasible
11	90218	Infeasible	58848	Infeasible	7.20E+06	Infeasible	9.70E+07	Infeasible	*	*	E+09	Infeasible
12	36204	Feasible	43794	Feasible	1.4 E+06	Feasible	4.50E+07	Feasible	*	*	E+09	Feasible
13	106082	Infeasible	111431	Infeasible	3.90E+06	Infeasible	1.20E+08	Infeasible	*	*	E+09	Infeasible
14	33699	Infeasible	32598	Infeasible	2.2 E+06	Infeasible	1.10E+08	Infeasible	*	*	E+09	Infeasible
15	64368	Infeasible	41837	Infeasible	3.80E+06	Infeasible	1.10E+08	Infeasible	*	*	E+09	Infeasible
16	38577	Feasible	18049	Feasible	1.9 E+06	Feasible	1.60E+08	Feasible	*	*	E+09	Feasible
17	26107	Feasible	7903	Feasible	1.2 E+06	Feasible	1.80E+07	Feasible	*	*	6.85E+08	Feasible
18	75633	Feasible	4723	Feasible	4.40E+06	Feasible	7.90E+07	Feasible	*	*	1.02E+09	Feasible
19	86061	Infeasible	48813	Infeasible	4.50E+06	Infeasible	1.30E+08	Infeasible	*	*	E+09	Infeasible
20	36737	Infeasible	43710	Infeasible	2.20E+06	Infeasible	1.40E+08	Infeasible	*	*	E+09	Infeasible
21	67555	Infeasible	46083	Infeasible	5.70E+06	Infeasible	2.00E+08	Infeasible	*	*	E+09	Infeasible
22	16170	Feasible	58411	Feasible	5.1 E+05	Feasible	2.90E+07	Feasible	*	*	E+09	Feasible
23	33843	Feasible	11267	Feasible	2.20E+06	Feasible	4.80E+07	Feasible	*	*	3.39E+08	Feasible
24	78172	Infeasible	40769	Infeasible	3.70E+06	Infeasible	1.30E+08	Infeasible	*	*	E+09	Infeasible
25	75375	Infeasible	74582	Infeasible	3.40E+06	Infeasible	2.40E+08	Infeasible	*	*	E+09	Infeasible

Therefore, if one solve whether there exists an integer point in the polytope (2) by the distributed implementation, one can settle the problem (1).

In this numerical results, 3 computers which are OptiPlex 330 model with 2 processors form the distributed network. All programs are in C++ language, and CPLEX Convert Technology is brought in to the linear programs in the fixed-point approach in this network. Each subpolytope which is segmented by the master is independent to each other. Therefore, each slave computer can carry out each subpolytope at the same time. Message Passing Interface [9] is introduced to build a communication network between the slaves and master. p equals 5. q equals 40. The other settings and parameters are the same as the ones in [12].

Given an example, 2 cases have been considered. The one is solving the system 1 directly. The other is solving the system 1 by computing the system 2. Three other methods, including branch and cut method, branch and bound method, Cplex constraint programming, have also been used to make comparison. The numerical results can be obtained in the following Fig. 1. In order to present numerical results simply, some symbols are used.

NumLPs: Iterations' number for a certain algorithm

F: Whether the example feasible or not

BC: The branch and cut method

BB: The branch and bound method

CP: Cplex constraint programming method

From the Table 1, one can see that this distributed method is the best one and the problem can be easier after the basis reduction. For a certain problem, the distributed method has the least number of iterations among these four algorithms. If the problem is solved without the basis reduction, it is harder. For the branch and bound algorithm, it is beyond its ability to solve the problem of this dimension.

4   Conclusion

In this paper, the implementation of distributed fixed-point approach has been built. The details and technologies of the implementation have been discussed. The market split problem is solved by this distributed implementation and other integer programming algorithms. The numerical results show that this fixed-point algorithm is very promising and can be used to other similar integer problems.

After this paper, one can see there are two purposes of this paper. One is to bring in the distributed method into some wicked integer problems and give the details of the distributed implementation. As it is very hard to solve some wicked problems, especially large-scale problems and algorithms for big data. Distributed method is a new trend of solving the problems of this kind. The other purpose of this paper is to share new numerical results of the market split problem using distributed of fixed-point method. Compared with branch and cut method, branch and bound method and Cplex constraint programming method, one can see this distributed of fixed-point method obtained the best. As known, this new numerical result of the market split problem using distributed of fixed-point method is the best one so far.

Acknowledgment. The authors are very grateful to the reviewers for their valuable suggestions and comments. This work was partially supported by National Nature Science Foundation of China under grants 71471091 and 71271119, Research Foundation of USTS under grants No. XKQ201517.

References

1. Gary, M.R., Johnson, D.S.: Computers and Intractability: A Guide to the Theory of NP-completeness. WH Freeman & Co., San Francisco (1979)
2. Gomory, R.E.: Outline of an algorithm for integer solutions to linear programs. Bull. Am. Math. Soc. **64**(5), 275–278 (1958)
3. Land, A.H., Doig, A.G.: An automatic method of solving discrete programming problems. Econometrica: J. Econometric Soc. **28**, 497–520 (1960)
4. Scarf, H.E.: Neighborhood systems for production sets with indivisibilities. Econometrica: J. Econometric Soc. **54**, 507–532 (1986)
5. Jünger, M., Liebling, T., Naddef, D., Nemhauser, G., Pulleyblank, W., Rerhard, G., Rinaldi, G., Wolsey, L.: 50 Years of Integer Programming 1958–2008. Springer, Berlin (2010)
6. Dang, C.: An increasing-mapping approach to integer programming based on lexicographic ordering and linear programming. In: The Ninth International Symposium on Operations Research and Its Applications, Chengdu-jiuzhaigou, China (2010)
7. Dang, C., Ye, Y.: A fixed point iterative approach to integer programming and its distributed computation. Fixed Point Theor. Appl. **2015**(1), 1–15 (2015)
8. Dantzig, G.B.: Linear Programming and Extensions. Princeton University Press, Princeton (1998)
9. Message Passing Interface Forum: MPI: A message-passing interfacestandard, version 2.2 (2009). http://www.mpiforum.org/docs/mpi-2.2/mpi22-report.pdf
10. Cornuéjols, G., Dawande, M.: A class of hard small 01 programs. In: Integer Programming and Combinatorial Optimization, pp. 284–293 (1998)
11. Aardal, K., Bixby, R.E., Hurkens, C.A., Lenstra, A.K., Smeltink, J.W.: Market split and basis reduction: towards a solution of the Cornuéjols-Dawande instances. INFORMS J. Comput. **12**(3), 192–202 (2000)
12. Wu, Z., Dang, C., Zhu, C.: Solving the market split problem using a distributed computation approach. In: IEEE International Conference on Information and Automation, Yinchuan, China, pp. 1252–1257 (2013)

Optimization of Supplier Selection and Order Allocation Under Fuzzy Demand in Fuzzy Lead Time

Sirin Suprasongsin[1](✉), Van Nam Huynh[1], and Pisal Yenradee[2]

[1] Japan Advanced Institute of Science and Technology, Nomi, Ishikawa, Japan
{sirin,huynh}@jaist.ac.jp
[2] Sirindhorn International Institute of Technology, Thammasat University,
Pathumthani, Thailand
pisal@siit.tu.ac.th

Abstract. This paper deals with the problem of Supplier Selection and Order Allocation (SSOA) in a fuzzy sense. The demand and delivery lead time are treated as fuzzy numbers. The fuzzy number is first transformed into interval numbers. After doing some arithmetic operations, those fuzzy interval numbers are defuzzified to a crisp quantity. This crisp quantity is further used as an input parameter in the model. Essentially, the main approach in this paper is based on the function principle and the pascal triangular graded mean approach. The SSOA problem is constructed as a Multiple Criteria Decision Making (MCDM) problem aiming to optimize the order quantities placed to many suppliers. The problem is solved by a fuzzy linear programming technique. A numerical example is also given for the illustration of the discussed issues.

Keywords: Function principle · Pascal triangular graded mean approach · Multiple-objective linear programming · Supplier Selection and Order Allocation problem

1 Introduction

Supplier selection is one of the classical problems of supply chain management. The decision of selecting suppliers has a long term impact on a whole supply chain. As observed by Ghodspour and O'Brien (2001) [1], 70 % of product cost comes from raw material costs. Hence, selecting appropriate suppliers may reduce the total cost of the supply chain. Furthermore, in recent trend, decision makers are more demanding, not only regarding the total cost, but also other satisfactions such as from a high product quality and from a precision of scheduled inventory. Therefore, for increasing competitive market, it is important for firms to have a good decision support approach reflecting their desire. Since there are many concerned criteria or objectives, the problem is usually treated as the Multiple Criteria Decision Making (MCDM) problem. MCDM problem has actually received much attention since decades. In addition, MCDM problem

© Springer Nature Singapore Pte Ltd. 2016
J. Chen et al. (Eds.): KSS 2016, CCIS 660, pp. 182–195, 2016.
DOI: 10.1007/978-981-10-2857-1_16

also frequently deals with information represented in qualitative and quantitative forms with uncertainty. So far, many integrated approaches have been introduced to deal with a MCDM problem in an uncertain environment. The mathematical model of fuzzy concepts for handling uncertainties was originally proposed by Zadeh (1965) [2] Since then, many applications of fuzzy set theory have been developed.

In most of the real-life supplier selection and order allocation problem, demand and delivery lead time are frequently associated with uncertainties and treated as fuzzy data. Practically, there are many situations that both demand and delivery lead time uncertainties occur simultaneously. This kind of situations essentially affects an inventory strategy. The inventory strategy is a systematic process for managing stocks which are important for any other supply chain departments. A periodic review inventory model is one of the classical inventory-review systems which is made on periodic basis such as every 15 days or 30 days. In this matter, the order quantities are placed following the review of on hand inventory.

Over the past decades, the application of fuzzy sets in economics order quantity (EOQ) has been developed by many researchers. Various purposes of production inventory model are optimized such as finding the optimal safety stock, reducing the total cost, finding the optimal reorder point, and so on. In 1987, Park [4] proposed an inventory model using fuzzy set concept under arithmetic operations of extension principle. In 1994, Tersine [3] proposed four inventory models with different conditions: stochastic demand and lead time, variable demand and lead time, stochastic demand and variable lead time, variable demand and stochastic lead time. Lee et al. (1998) [5] developed a fuzzy inventory model under fuzzy demand and fuzzy production quantity. Hsieh (2002, 2004) [6,7] proposed a model to find the optimal reorder point and optimal safety stock under fuzzy demand and fuzzy lead time. More recently, demand and lead time uncertainties have received much interest from research community. Mahata and Goswami (2009) [8] determined economic order quantity by taking demand and lead time as a fuzzy data. Taleizadeh et al. (2012) [9] discussed a multiple products multiple suppliers in supply chain with stochastic demand and variable lead time. Later, in 2013, Taleizadeh et al. [10] developed a hybrid method of fuzzy simulation and genetic algorithm for inventory control systems with stochastic replenishments and fuzzy demand. Sakar and Majumder (2013) [11] introduced an inventory model with variable time. The objective of the model was to reduce a set up cost. Sakar and Moon (2014) [12] considered variable lead time and variable reorder point with a fixed demand. In 2015, Sakar and Mahapatra [13] proposed a fuzzy inventory model with variable lead time, variable reorder point and fuzzy demand. The main goal of this paper was to reduce the total annual cost. See a comparison between our study and other research works in Table 1.

To deal with a problem of the inter-relation fuzzy data, various methods for operating fuzzy numbers have previously been proposed in the literatures. Extension principle is one of the major techniques for operating fuzzy number as

Table 1. A comparison between our study and other research works

Year	Author(s)	Fuzzy demand	Fuzzy lead time	Fuzzy * Fuzzy	Multiple criteria
1987	Park	No	No	No	No
1998	Lee and Yao	Yes	No	No	No
2002	Hsieh	Yes	No	Yes	No
2004	Hsieh	Yes	Yes	Yes	No
2009	Mahata and Goswami	Yes	Yes	No	Yes
2011	Taleizadeh et al.	Yes	No	No	No
2013	Taleizadeh et al.	Yes	No	No	No
2013	Sakar and Majumder	No	No	No	No
2014	Sakar and Moon	No	No	No	No
2015	Sakar and Mahapatra	Yes	Yes	No	No
-	**This model**	Yes	Yes	Yes	Yes

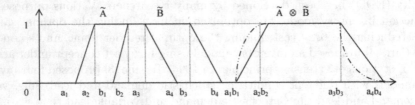

Fig. 1. The fuzzy multiplication operation of function principle and extension principle

mentioned by Zadeh (1975) [14]. However, the computation under the extension principle is very complicated. Therefore, in 1985, Chen et al. [15] introduced the function principle to ease the computation of the extension principle. The principle is proven that it does not change the type of membership function under fuzzy arithmetical operations of fuzzy number. To illustrate more, Hsieh [6] compared these concept as shown in Fig. 1.

Moreover, in a decision process, most of the cases naturally require a crisp value as a solution, instead of a fuzzy value, because of the limitation of human perception. Hence, a defuzzifying process is required to convert fuzzy quantities into crisp quantities. With the recognition of defuzzifying process, many advanced techniques have been developed such as adaptive integration (AI), center of gravity (COG), fuzzy clustering defuzzification (FCD), first of maximum (FOM), weighted fuzzy mean (WFM), graded mean integration representation, and pascal triangular graded mean approach.

Motivated by the above observation that there is no consideration of fuzzy demand in fuzzy lead time existed in literatures in the SSOA problem, we develop a model considering fuzzy demand in fuzzy lead time. In this paper, we use function principle and pascal triangular graded mean approach to find the percent error of forecasted demand which is affected by discount policy from suppliers. The percent error is calculated from the error of fuzzy demand in fuzzy lead time compared with the most likely value (benchmarking). Then, this percent error is treated as a criterion in the SSOA decision model so that the optimal order quantities placed to suppliers under practical constraints can be determined.

The organization of this paper is as follows. In Sect. 2, the methodologies in determining a percent error of demand are described step by step. In Sect. 3, fuzzy programming technique is illustrated. Then, model development is introduced in Sect. 4. Remarks and Conclusion are presented in Sect. 5.

2   How to Defuzzify Fuzzy Demand in Fuzzy Lead Time

Fuzzification is a process of changing the scalar value into the fuzzy value described by different types of membership functions. We do this based on the recognition that many quantities carry considerable uncertainty. Generally, uncertainty arises from the imprecision, vagueness and ambiguity. Fuzzy values are used as the input structure for a fuzzy system. This idea is shown in Fig. 2, where we consider demand as fuzzy demand.

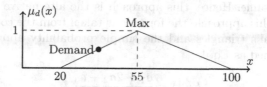

Fig. 2. Triangular membership function representing imprecision in demand

In this paper, we use function principle and pascal triangular graded mean approach to defuzzify fuzzy demand in fuzzy lead time. Essentially, pascal triangular graded mean approach is an alternative approach extended from Graded mean integration representation approach. To do so, four relevant methodologies are introduced as follows.

2.1   Graded Mean Integration Representation Approach

In 1998, Chen and Hsieh [16] proposed graded mean Integration representation approach for defuzzifying generalized fuzzy numbers. As indicated by Babu and Anand [17], this method is useful to obtain the solution of the generalized fuzzy variables for assignment problem.

Suppose L^{-1} and R^{-1} are inverse functions of function L and R, respectively and the graded mean h-level of generalized fuzzy number $A = (a_1, a_2, a_3 : w)$ is

$\frac{h[L^{-1}(h)+R^{-1}]}{2}$. Then the defuzzified value P(A) based on the integral value of graded mean h-level can be defined using Eq. 1.

$$P(A) = \frac{\int_0^h [\frac{L^{-1}(h) + R^{-1}(h)}{2}]dh}{\int_0^w h\,dh} \tag{1}$$

where h is in between 0 and w, $0 < w \leq 1$. Interestingly, the representation of fuzzy number can be generalized as shown in Eqs. 2 and 3. For example, if $A = (a_1, a_2, a_3)$ is a triangular fuzzy number, then

$$P(A) = \frac{1}{2}\frac{\int_0^1 \int h[a_1 + h(a_2 - a_1) - h(a_3 - a_2)]dh}{\int_0^1 h\,dh} \tag{2}$$

$$P(A) = \frac{a_1 + 4a_2 + a_3}{6} \tag{3}$$

2.2 Pascal Triangular Graded Mean Approach

Pascal triangular graded mean approach is an extension of Graded mean integration representation approach. It is applicable for multiple-objective fuzzy assignment problem. In 2013, Babu and Anand [17] indicated that there is no significant difference between Graded mean integration representation approach and pascal triangular graded mean approach. From the statistical test, mean and variance are the same. Hence, this approach is the alternative approach of the previous one. In this approach, the formula is taken from the coefficient of fuzzy numbers of pascal's triangles and the simple probability approach. Then the formula is simplified as Eq. 4.

$$P(A) = \frac{a_1 + 2a_2 + a_3}{4} \tag{4}$$

2.3 Fuzzy Arithmetical Operations Under Function Principle

In order to simplify the calculation of extension principle, Chen (1985) [18] introduced function principle to operate fuzzy numbers. Some fuzzy arithmetical operations under function principle are described below.

Suppose $\tilde{A} = (a_1, a_2, a_3)$ and $\tilde{B} = (b_1, b_2, b_3)$ are two sets of triangular fuzzy numbers. Then the followings are four arithmetical operations that can perform on triangular fuzzy numbers.

1. The addition of $\tilde{A}$ and $\tilde{B}$
   $\tilde{A} + \tilde{B} = (a_1, a_2, a_3) + (b_1, b_2, b_3) = (a_1 + b_1, a_2 + b_2, a_3 + b_3)$
2. The subtraction of $\tilde{A}$ and $\tilde{B}$
   $\tilde{A} - \tilde{B} = (a_1, a_2, a_3) - (b_1, b_2, b_3) = (a_1 - b_1, a_2 - b_2, a_3 - b_3)$
3. The multiplication of $\tilde{A}$ and $\tilde{B}$ is $\tilde{A} \times \tilde{B} = (c_1, c_2, c_3)$ where $T = a_1b_2, a_1b_3, a_3b_1, a_3b_3$ $c_1 = \min T$, $c_2 = a_2b_2$, $c_3 = \max T$
   However, if $a_1, a_2, a_3, b_1, b_2, b_3$ are non-zero positive real numbers, then
   $\tilde{A} \times \tilde{B} = (a_1, a_2, a_3) \times (b_1, b_2, b_3) = (a_1b_1, a_2b_2, a_3b_3)$

4. The division of $\tilde{A}$ and $\tilde{B}$ is $\frac{\tilde{A}}{\tilde{B}} = (c_1, c_2, c_3)$ where $T = \frac{a_1}{b_2}, \frac{a_1}{b_3}, \frac{a_3}{b_1}, \frac{a_3}{b_3}$
   $c_1 = \min T$, $c_2 = \frac{a_2}{b_2}$, $c_3 = \max T$
   However, if $a_1, a_2, a_3, b_1, b_2, b_3$ are non-zero positive real numbers, then
   $\frac{\tilde{A}}{\tilde{B}} = (a_1, a_2, a_3) \div (b_1, b_2, b_3) = (\frac{a_1}{b_1}, \frac{a_2}{b_2}, \frac{a_3}{b_3})$

2.4 Determine the Percent Error of Fuzzy Expected Demand (Fuzzy Demand in Fuzzy Lead Time $\tilde{d} \times \tilde{l}$)

According to the theories of function principal and pascal triangular graded mean approach, fuzzy expected demand or fuzzy demand in fuzzy lead time can be determined. Fuzzy expected demand results from a multiplication of fuzzy demand and fuzzy lead time. In this paper, fuzzy expected demand represents the demand in thirty days concerning with uncertainties in forecasted demand and delivery lead time. Steps to determine the percent error of fuzzy expected demand are as follows.

1. Multiply fuzzy demand $\tilde{d}$ (unit/day) and fuzzy lead time $\tilde{l}$ (day) from Tables 2 and 3. The results are determined based on the concept of fuzzy arithmetical operation defined by function principle. Note that the concerned fuzzzy numbers are non-zero positive real numbers. The results are called fuzzy expected demand (a monthly demand) and presented in Table 4.
   Ex. From daily demand of Product 1 which is supplied by Supplier 1,
   $\tilde{d} \times \tilde{l} = (d_1, d_2, d_3) \times (l_1, l_2, l_3) = (d_1 l_1, d_2 l_2, d_3 l_3)$
   $\tilde{d} \times \tilde{l} = (233, 333, 433) \times (25, 30, 32) = (5833, 10000, 13867)$
2. With the use of pascal triangular graded mean approach (Eq. 4), Fuzzy expected demand per month is defuzzified as shown in Table 5.
   Ex. Pascal defuzzification of Product 1 supplied by Supplier 1,
   $P(A) = \frac{a_1 + 2a_2 + a_3}{4}$
   $P(A) = \frac{5833 + 2(10000) + 13867}{4} = 9925$
3. The percent error of fuzzy expected demand and the forecasted demand is determined in Table 6. From now on, the forecasted demand shown in Table 7 is named as the benchmarking. Note that the percent errors are used as the input parameters for a further optimization process.
   Ex. % error $= |(\text{Forecasted demand} - \text{Expected demand})| \div (\text{Forecasted demand})$
   % error $= \frac{10000 - 9925}{10000} = 0.75$

3 Fuzzy Programming Technique

To solve the multiple-objective decision making problem, we use the fuzzy programming technique. For each of objective functions, we first find the upper bound mx_i and the lower bound mn_i, where mx_i and mn_i are the greatest value of criteria i and the lowest value of criteria i, respectively. Once the boundaries have been found, we then optimize the fuzzy model. The steps of fuzzy multiple-objective linear programming technique are as follows.

1. Solve the multiple-objective functions as a single objective at a time. For example, if there are totally three objectives, three optimizations have to be done. The result is exemplified in Table 8.
2. From Table 8, we determine the upper bound and the lower bound for each objective as presented in Table 9. Then these values are used as input parameter in Step 4.
3. Determine the linear membership functions for each objective function. For example, if a fuzzy number A is a triangular fuzzy number defined by (a_1, a_2, a_3), then the membership function is defined by

$$\mu_A(x) = \begin{cases} 1 & \text{if } x \geq a_3 \\ \frac{x-a_1}{a_2-a_1} & \text{if } a_1 \leq x \leq a_2 \\ 0 & \text{if } x \leq a_1 \end{cases}$$

$$\mu_A(x) = \begin{cases} 1 & \text{if } x \leq a_1 \\ \frac{a_3-x}{a_3-a_2} & \text{if } a_2 \leq x \leq a_3 \\ 0 & \text{if } x \geq a_3 \end{cases}$$

where the first one is used for maximizing objective and the latter is used for minimizing objective.
4. Determine the optimal solution of the multiple-objective linear programming problem as illustrated in Sect. 4.

4   Model Development

We develop a model for supplier selection and order allocation problem under price discount environment. The proposed mathematical model is the fuzzy multiple-objective linear programming.

4.1   Problem Description

This problem is a multiple criteria decision making problem with uncertainty through a tutorial example taken from [19]. In this example, a decision maker must properly allocate the proper order quantities of products to suppliers based on given constraints. Let us further illustrate the problem as follows.

- There are 5 products and 5 suppliers under consideration.
- Supplier k $(k = 1, ..., K)$ offers either volume discount or quantity discount when product j $(j = 1, ..., J)$ is purchased at a discount level c $(c = 1, ..., C)$. Note that in this case, only supplier 3 offers a volume discount policy.
- 3 criteria are concerned, namely, total cost, quality, and % error of fuzzy expected demand. The % error of fuzzy expected demand is measured from the percent error of fuzzy demand in fuzzy lead time compared to the total monthly forecasted demand as described in Sect. 2. In addition, relative importances of criteria (weights) are given.

- 3 objectives are considered, namely, (a) minimizing the total cost under volume and quantity discount constraint, (b) maximizing quality, and (c) minimizing % error of fuzzy expected demand.
- Satisfaction level which is dimensionless is used to transform various units of criteria
- Assume that all suppliers supply products once a month.
- Forecasted demand is uncertain due to the price discount offered by suppliers to spur up their sales.

4.2   Notations

In order to formulate the model, the following notations are defined.

Indices

i index of criteria	$i = 1...I$
j index of products	$j = 1...J$
k index of suppliers	$k = 1...K$
c index of price-break levels	$c = 1...C$

Input parameters

dc_j constant (crisp) demand of product j (unit)

h_{jk} capacity for product j from supplier k (unit)

p_{cjk} price of product j offered from supplier k at discount level c (\$)

$z1_{jk}$ price of product j per unit offered by supplier k (\$)

$z2_{jk}$ quality score of product j from supplier k (scores)

$z3_{jk}$ percent error of fuzzy expected demand (%)

e_{cjk} price-break level c of product j from supplier k for quantity discount (unit)

b_{ck} price-break level c from supplier k for volume discount(\$)

g_{ck} percent volume discount from supplier k at level c (unitless)

f_k 1 if supplier k offers quantity discount; 0 otherwise (unitless)

w_i weight of criteria i (unitless)

mn_i minimum value of criteria i (Lower bound)(\$, scores, %)

mx_i maximum value of criteria i (Upper bound)(\$, scores, %)

Decision variables

x_{cjk} purchased quantity at discount level c of product j from supplier k (unit) at constant demand

π_{jk} 1 if supplier k supplies product j; 0 otherwise (unitless) fuzzy expected demand (unitless)

t_{cjk} total purchasing cost j from supplier k at level c for quantity discount (\$)

a_{ck} total purchasing cost from supplier k at level c for volume discount (\$)

α_{ck} 1 if quantity discount level c is selected for supplier k; 0 otherwise (unitless)

β_{ck} 1 if volume discount level c is selected for supplier k; 0 otherwise (unitless)

λ_i satisfaction levels of criteria i; cost, quality and percent error of fuzzy expected demand (unitless)

4.3   Mathematical Formulation

In this study, for the sake of simplicity, weighted additive model is applied. A basic concept of this model is to use a single utility function representing the overall preference of DMs corresponding to the relative importance of each criterion.

Maximize

$$\Sigma_i w_i \cdot \lambda_i \tag{5}$$

Price discount. Quantity discount and volume discount are considered in this model. Constraints (6–11) indicate that purchasing quantity x_{cjk} must be corresponding to a suitable discount level for both discount policies. In addition, only one level is selected. The relevant data are shown in Tables 10, 11 and 12.

$$\Sigma_c t_{cjk} \cdot f_k = \Sigma_c p_{cjk} \cdot x_{cjk} \cdot f_k \qquad \forall j, k \tag{6}$$

$$e_{c-1,jk} \cdot \alpha_{ck} \cdot f_k \leq \Sigma_j x_{cjk} \cdot f_k < e_{cjk} \cdot \alpha_{ck} \cdot f_k \qquad \forall c, k \tag{7}$$

$$\Sigma_c \alpha_{ck} \cdot f_k \leq 1 \qquad \forall k \tag{8}$$

$$\Sigma_c a_{ck} \cdot (1 - f_k) = \Sigma_c \Sigma_j z1_{jk} \cdot x_{cjk} \cdot (1 - f_k) \qquad \forall k \tag{9}$$

$$b_{c-1,k} \cdot \beta_{ck} \cdot (1 - f_k) \leq a_{ck} \cdot (1 - f_k) < b_{ck} \cdot \beta_{ck} \cdot (1 - f_k) \qquad \forall c, j, k \tag{10}$$

$$\Sigma_c \beta_{ck} \cdot (1 - f_k) \leq 1 \qquad \forall k \tag{11}$$

Capacity. The total purchasing quantity x_{cjkn} can not be greater than supply capacity h_{jk} and it is considered only when the assigned π_{jk} is equal to 1. Numerical data are presented in Table 13.

$$\Sigma_c x_{cjk} \leq h_{jk} \cdot \pi_{jk} \qquad \forall j, k \tag{12}$$

Satisfaction level. Constraints (13–16) describe the satisfaction levels of cost, quality, and % error of fuzzy expected demand.

$$\lambda_1 \leq \frac{mx_1 - \Sigma_c \Sigma_j \Sigma_k t_{cjk} \cdot f_k + \Sigma_c \Sigma_k a_{ck} \cdot (1 - g_{ck}) \cdot (1 - f_k)}{mx_1 - mn_1} \tag{13}$$

$$\lambda_2 \leq \frac{\Sigma_c \Sigma_j \Sigma_k z2_{jk} \cdot x_{cjk} - mn_2}{mx_2 - mn_2} \tag{14}$$

$$\lambda_3 \leq \frac{mx_3 - \Sigma_c \Sigma_j \Sigma_k z3_{jk} \cdot x_{cjk}}{mx_3 - mn_3} \tag{15}$$

$$0 \leq \lambda_i < 1 \qquad \forall i \tag{16}$$

Consequently, optimal purchasing quantities are shown in Table 14. In addition, from Table 15, we also see that the result reflects very well with the preference of decision makers. The ranking of given criteria weights and satisfaction levels is the same, ranging from cost > % error of fuzzy expected demand > quality criteria.

Table 2. Fuzzy demand per day $\tilde{d}$ (unit/day)

Level/Product	P1	P2	P3	P4	P5
Minimum variation	233	200	117	83	77
Predicted demand	333	267	200	100	83
Maximum variation	433	300	267	167	100

Table 3. Fuzzy lead time $\tilde{l}$ (day)

Level/Product	S1	S2	S3	S4	S5
Minimum variation	25	30	28	26	23
Predicted demand	30	30	30	30	30
Maximum variation	32	30	31	35	37

Table 4. Fuzzy expected demand per month based on function principle $\tilde{d} \times \tilde{l}$ of Supplier 1

Level/Product	P1	P2	P3	P4	P5
Minimum variation	5833	5000	2917	2083	1917
Predicted demand	10000	8000	6000	3000	2500
Maximum variation	13867	9600	8533	5333	3200

Table 5. Pascal defuzzification

Product/Supplier	S1	S2	S3	S4	S5
P1	9925	7650	5863	3354	2529
P2	10000	7750	5875	3375	2575
P3	9992	7725	5883	3375	2562
P4	10308	7925	6092	3500	2623
P5	10350	7925	6138	3521	2616

- Given weights of cost, % error of fuzzy expected demand, and quality criteria are 0.5, 0.32, and 0.18, respectively.
- Optimal satisfaction levels of cost, % error of fuzzy expected demand, and quality criteria are 1, 0.69, and 0.64, respectively.

The results then can support decision makers to decide the appropriate order quantities placed to suppliers.

Table 6. Unit (LIST) price, Quality score and % error of fuzzy expected demand; $(z1_{jk})$, $(z2_{jk})$ and $(z3_{jk})$

Data	Product/Supplier	S1	S2	S3	S4	S5
Unit (List) price	P1	50	40	55	50	45
	P2	0	200	0	230	0
	P3	70	75	69	0	0
	P4	0	0	30	32	29
	P5	0	0	0	19	20
Quality score	P1	3	8	6	2	4
	P2	0	6	0	7	0
	P3	5	7	8	0	0
	P4	0	0	8	10	5
	P5	0	0	0	8	9
% error of fuzzy expected demand	P1	0.75	4.38	2.29	11.80	1.17
	P2	0.00	3.13	2.08	12.50	3.00
	P3	0.08	3.44	1.94	12.50	2.47
	P4	3.08	0.94	1.53	16.67	4.93
	P5	3.50	0.94	2.29	17.37	4.63

Table 7. Forecast demand of each product (Benchmarking) (dc_j)

Product	Forecast demand
1	10000
2	8000
3	6000
4	3000
5	2500

Table 8. Result from three optimizations

Criteria i	Lowest value	Medium value	Highest value
z1 (Cost)	2467637	2738000	2533558
z2 (Quality score)	194413	231388	154514
z3 (% error of fuzzy expected demand)	158454	220005	116336

Table 9. Boundaries of each criterion (mn_i, mx_i)

Criteria i	mn_i	mx_i	Units
z1 (Cost)	2467637	2738000	$
z2 (Quality score)	154514	231388	Score
z3 (% error of fuzzy expected demand)	116336	220005	%

Table 10. Break point of quantity discount at level (e_{cjk})

Product/Supplier	S1	S2	S4	S5
P1	5000	7000	6000	8000
P2	4000	2000	3000	2500
P3	4000	4000	2500	3500
P4	1200	3000	1500	2000
P5	1200	3000	1500	2000

Table 11. Price of each product for quantity discount levels (p_{cjk})

Level/Sup.	S1			S2				S4					S5			
	P1	P3	P2,4,5	P1	P2	P3	P4-5	P1	P2	P3	P4	P5	P1	P2-3	P4	P5
Level 1	50	70	0	40	200	75	0	50	230	0	32	19	45	0	29	20
Level 2	45	68	0	39	180	74	0	48	220	0	30	18	43	0	28	17
Level 3	43	65	0	38	170	73	0	46	210	0	28	16	42	0	25	14

Table 12. Break point of volume discount (b_{ck}) and volume discount percentage (g_{ck})

Level	Supplier 3	
	b_{ck}	g_{ck}
1	0	0
2	500000	0.05
3	2000000	0.1

Table 13. Capacity (h_{jk})

Product/Supplier	S1	S2	S3	S4	S5
P1	10000	8000	6000	30000	5000
P2	0	6000	0	10000	0
P3	5000	4000	6000	0	0
P4	0	0	8000	10000	5000
P5	0	0	0	60000	2000

Table 14. Optimal purchasing quantity from weighted additive technique

Product/Supplier	S1	S2	S3	S4	S5
P1	-	6651	-	3346	3
P2	-	5999	-	2001	-
P3	-	-	6000	-	-
P4	-	-	3000	-	-
P5	-	-	-	653	1847

Table 15. Weight sets (w_i) and Optimal satisfaction level (λ_i)

Criteria	Weight	Optimal satisfaction level
Cost	0.5	1.00
% error	0.32	0.69
Quality	0.18	0.64

5   Remarks and Conclusion

In this paper, a concept of multiplying two fuzzy numbers by arithmetic fuzzy operation called function principle and a concept of defuzzification by pascal triangular graded mean approach are first introduced to a Supplier Selection and Order Allocation (SSOA) problem. The function principle is used to simplify the calculation of fuzzy expected demand derived from the multiplication of fuzzy demand and fuzzy delivery lead time. Then, it is defuzzified by pascal triangular graded mean approach. Consequently, fuzzy demand is used as an input parameter of a criterion named a percent error of fuzzy expected demand. The numerical example is demonstrated. With the use of fuzzy programming technique, we have derived the solution of multiple-objective deterministic model. This method is very simple while comparing to other methods.

References

1. Ghodsypour, S.H., O'Brien, C.: The total cost of logistics in supplier selection, under conditions of multiple sourcing multiple criteria and capacity constraint. Int. J. Prod. Econ. **73**(1), 15–27 (2001)
2. Zadeh, L.A.: Fuzzy sets. Inf. Control **8**(3), 338–353 (1965)
3. Tersine, R.J.: Principles of Inventory and Materials Management. Prentice Hall, Englewood Cliffs (1994)
4. Park, K.S.: Fuzzy-set theoretic interpretation of economic order quantity. IEEE Trans. Syst. Man Cybern. **17**(6), 1082–1084 (1987)
5. Lee, H.M., Yao, J.S.: Economic production quantity for fuzzy demand quantity, and fuzzy production quantity. Eur. J. Oper. Res. **109**(1), 203–211 (1998)
6. Hsieh, C.H.: Optimization of fuzzy production inventory models. Inf. Sci. **146**(1), 29–40 (2002)
7. Hsieh, C.H.: Optimization of fuzzy inventory models under fuzzy demand and fuzzy lead time. Tamsui Oxf. J. Manage. Sci. **20**(2), 21–36 (2004)
8. Mahata, G.C., Goswami, A.: An EOQ model with fuzzy lead time, fuzzy demand and fuzzy cost coefficients. Int. J. Eng. Appl. Sci. **5**(5), 295–302 (2009)
9. Taleizadeh, A.A., Niaki, S.T.A., Makui, A.: Multiproduct multiple-buyer single-vendor supply chain problem with stochastic demand, variable lead-time, and multi-chance constraint. Expert Syst. Appl. **39**(5), 5338–5348 (2012)
10. Taleizadeh, A.A., Niaki, S.T.A., Aryanezhad, M.B., Nima, S.: A hybrid method of fuzzy simulation and genetic algorithm to optimize constrained inventory control systems with stochastic replenishments and fuzzy demand. Inf. Sci. **220**, 425–441 (2013)

11. Sakar, B., Majumder, A.: Integrated vendor-buyer supply chain model with vendors setup cost reduction. Appl. Math. Comput. **224**, 362–371 (2013)
12. Sakar, B., Moon, I.: Improved quality, setup cost reduction, and variable backorder costs in an imperfect production process. Int. J. Prod. Econ. **155**, 204–213 (2014)
13. Sakar, B., Mahapatra, A.S.: Periodic review fuzzy inventory model with variable lead time and fuzzy demand. Int. Trans. Oper. Res. (2015)
14. Zadeh, L.A.: The concept of a linguistic variable and its application to approximate reasoning I. Inf. Sci. **8**(3), 199–249 (1975)
15. Chen, S.H.: Fuzzy linear combination of fuzzy linear functions under extension principle and second function principle. Tamsui Oxf. J. Manage. Sci. **1**, 11–31 (1985)
16. Chen, S.H., Hsieh, C.H.: Graded mean integration representation of generalized fuzzy numbers. Chin. Fuzzy Syst. Assoc. **5**(2), 1–7 (1999)
17. Babu, S.K., Anand, R.: Statistical optimization for generalised fuzzy number. Int. J. Mod. Eng. Res. **3**(2), 647–651 (2013)
18. Chen, S.H.: Operations on fuzzy numbers with function principal. J. Manage. Sci. **6**(1), 13–26 (1985)
19. Suprasongsin, S., Yenradee, P., Huynh, V.N.: Suitable aggregation operator for a realistic supplier selection model based on risk preference of decision maker. In: Torra, V., Narukawa, Y., Navarro-Arribas, G., Yañez, C. (eds.) MDAI 2016. LNCS (LNAI), vol. 9880, pp. 68–81. Springer, Heidelberg (2016). doi:10.1007/978-3-319-45656-0_6

A Dynamic Spectrum Allocation Strategy in CRNs and Its Performance Evaluation

Shunfu Jin[1] and Wuyi Yue[2(✉)]

[1] School of Information Science and Engineering, Yanshan University,
Qinhuangdao 066004, China
jsf@ysu.edu.cn
[2] Department of Intelligence and Informatics, Konan University,
Kobe 658-8501, Japan
yue@konan-u.ac.jp

Abstract. Aiming to investigate the influence of the burst requests and the likelihood of impatience possibility on a dynamic spectrum allocation strategy in Cognitive Radio Networks (CRNs), we establish a batch arrival queueing model with possible reneging and potential transmission interruption. We derive the stead-state distribution of the queueing length to analyze the stochastic behavior of the system. Accordingly, we give some important performance measures and numerical results to show the change trends of the performance measures with respect to different batch arrival rates and different impatience parameters.

Keywords: Cognitive radio networks · Dynamic spectrum strategy · Batch arrival · Impatient packets · Markov chain

1 Introduction

Recently, scholars have carried out research on CRNs in relation to dynamic spectrum allocation strategy and performance evaluation. In [1], considering that the primary user may interrupt the transmission of a secondary user packet, Li et al. built a kind of Naor's model for CRNs, and gave the individually optimal equilibrium strategy and socially optimal strategy. In [2], taking into account the variable transmission mode and the fixed transmission mode, respectively, Farraj et al. built an M/G/1 queueing model to analyze the average transmission rate, the average service time and the average waiting time of secondary user packets. In [3], combing the cognitive process and the communication process into a queueing framework, Oklander et al. constructed a continuous-time Markov chain model. By means of a matrix geometric solution, they obtained the average delay and the throughput. In [4], Cadeau et al. constructed a Markov chain to capture the operations of spectrum sensing, channel accessing and channel switching under jamming. They also gave the closed-form expressions of the jamming probability and the throughput with various jamming models. In [5], considering a finite capacity, impatient secondary users, individual arrivals and

© Springer Nature Singapore Pte Ltd. 2016
J. Chen et al. (Eds.): KSS 2016, CCIS 660, pp. 196–202, 2016.
DOI: 10.1007/978-981-10-2857-1_17

channel reservations, Wang et al. constructed a system model for the session-level performance analysis in multi-channel CRNs. They also developed a multi-dimensional continuous-time Markov chain to obtain the dropping probability and the balk probability of the secondary users.

However, in the research works mentioned above, the data requests of the secondary user were generally assumed to arrive individually, especially for the systems with impatient secondary user packets. We know that in practical applications in CRNs, multiple secondary users may send out requests simultaneously, or one secondary user will send out burst requests. Also, a secondary packet waiting in the buffer may leave the system before the transmission occurs due to its impatience. In this paper, by considering both the burst arrival mode and impatience possibility of secondary user packets in CRNs with dynamic spectrum allocation, we construct a batch arrival queueing model to evaluate the performance measures of the network system.

2   System Model and Analysis

2.1   System Model for the Dynamic Spectrum Allocation Strategy

Dynamic spectrum allocation strategy is implemented in CRNs. During the transmission procedure of a secondary user packet, if a primary user packet arrives at the system, the primary user packet will take over the spectrum, while the secondary user packet being transmitted will terminate its transmission immediately and leave the system. The other secondary user packets in the buffer will be switched to another available spectrum.

In this paper, we consider an authorized spectrum in a centralized CRN in order to present the system model. We assume that there are always available spectrums to be allocated for the interrupted secondary users. Based on the spectrum condition in CRNs, the central controller will allocate one of the idle spectrums for the preempted secondary user packet. We suppose that there are always available spectrums for use. The dispatch process where the central controller allocates idle spectrums is not considered in this paper. Moreover, we assume that the secondary user packets in the buffer may leave the system due to their impatience while waiting in the buffer.

The centralized spectrum allocation strategy can be modeled as a batch arrival queueing model with possible reneging and potential transmission interruption.

We consider the system model in a continuous time field and assume that the system capacity is c $(0 < c < \infty)$. Referencing previous works (e.g. [4], [6], [7]), we make assumptions to develop our analytical model. We suppose that the batch arrival process of secondary user packets follows a Poisson distribution with parameter Λ $(\Lambda > 0)$, and the batch size follows a geometric distribution with parameter α $(0 < \alpha < 1)$. So the average number of secondary user packets arriving at the system per unit time is $\lambda_1 = \dfrac{\Lambda}{\alpha}$. We suppose that the time period required by a secondary user packet to be transmitted successfully follows an exponential distribution with parameter μ ($\mu > 0$). We also suppose that the maximum time for a secondary user packet waiting at the buffer before its transmission follows an exponential distribution with parameter β ($\beta > 0$). β is called the impatience parameter. Moreover, we suppose that the arrival process of the primary user packets follows another Poisson distribution with parameter λ_2 ($\lambda_2 > 0$).

2.2   Analysis in Steady State

Let $X(t)$ be the total number of the secondary user packets (including the secondary user packet occupying the spectrum) in the system at the time instant t ($t > 0$). With the assumptions given in Subsect. 2.1, the probability process $\{X(t), t > 0\}$ has the Markov property. If we call the number $X(t)$ of secondary user packets in the system "the system level", then the system is depicted by the system level $\{X(t) = 0, 1, 2, \ldots\}$. Let $Q(i, j)$ be the one step transition rate from system level i to j. $Q(i, j)$ can be illustrated as follows.

(1) When $i = 0$ and $j = 0$, $Q(0, 0)$ represents the probability that the number of secondary user packets being fixed at 0 after a one step transition. It means that no secondary user packet arrives at the system. So, $Q(0, 0) = -\Lambda$.
(2) When $j = i - 1$ ($1 \leq i \leq c$), $Q(i, i - 1)$ represents the probability that the number of secondary user packets in the system decreases by 1 after a one step transition. $Q(i, i - 1) = \mu + (i - 1)\beta + \lambda_2$.
(3) When $i = j$ ($1 \leq i \leq c-1$), $Q(i, i)$ represents the probability of the number of secondary user packets in the system being fixed at i after one step transition. $Q(i, i) = -\Lambda - (i - 1)\beta - \mu - \lambda_2$.
(4) When $0 \leq i \leq c - 1$ and $i + 1 \leq j \leq c - 1$, $Q(i, j)$ represents the probability that the number of secondary user packets in the system increases by k ($k = j - i$) after a one step transition. $Q(i, j) = \Lambda \alpha \bar{\alpha}^{k-1}$.
(5) When $0 \leq i \leq c - 1$ and $j = c$, $Q(i, c)$ represents the probability that the number of secondary user packets in the system increases to c after a one step transition. $Q(i, c) = \sum_{k=c}^{\infty} \Lambda \alpha \bar{\alpha}^{k-i} = \Lambda \bar{\alpha}^{c-i}$.
(6) When $i = c$ and $j = c$, $Q(c, c)$ represents the probability of the number of secondary user packets in the system being fixed at c after a one step transition. $Q(c, c) = -(i - 1)\beta - \mu - \lambda_2$.

In order to give the transition rate matrix $\boldsymbol{Q}$ of the system model more concisely, we introduce a notation $B = -(i - 1)\beta - \mu - \lambda_2$. Concluding the discussions above, the transition matrix $\boldsymbol{Q}$ with order $(c + 1) \times (c + 1)$ can be given as follows:

$$Q = \begin{pmatrix} -\Lambda & \Lambda\alpha & \Lambda\alpha\bar{\alpha} & \cdots & \Lambda\alpha\bar{\alpha}^{c-3} & \Lambda\alpha\bar{\alpha}^{c-2} & \Lambda\bar{\alpha}^{c-1} \\ -B & -\Lambda+B & \Lambda\alpha & \cdots & \Lambda\alpha\bar{\alpha}^{c-4} & \Lambda\alpha\bar{\alpha}^{c-3} & \Lambda\bar{\alpha}^{c-2} \\ & -B & -\Lambda+B & \cdots & \Lambda\alpha\bar{\alpha}^{c-5} & \Lambda\alpha\bar{\alpha}^{c-4} & \Lambda\bar{\alpha}^{c-3} \\ & & & \ddots & \vdots & \vdots & \vdots \\ \mathbf{0} & & & & -B & -\Lambda+B & \Lambda \\ & & & & & -B & B \end{pmatrix}. \tag{1}$$

Let π_i $(0 \le i \le c)$ be the stead-state probability vector for the system being at the level i. π_i can be given as follows:

$$\pi_i = \lim_{t\to\infty} P\{X(t) = i\}. \tag{2}$$

Let Π be the stead-state probability vector of the queueing length. Π can be given by

$$\Pi = (\pi_0, \pi_1, \ldots, \pi_c). \tag{3}$$

Combining the stead-state equation and the normalization condition, we have

$$\begin{cases} \Pi Q = 0 \\ \Pi e = 1 \end{cases} \tag{4}$$

where e is a column vector with $c+1$ elements, all of which equal 1.

By using the recursive method [6], we can calculate the stead-state distribution of the queueing length with numerical results.

3   Performance Measures

We define the loss rate p_l of secondary user packets as the average number of secondary user packets leaving the system due to their impatience or being interrupted by primary user packets per unit time. The loss rate p_l of secondary user packets can be given as follows:

$$p_l = \beta \sum_{i=0}^{c-1} i\pi_{i+1} + \lambda_2(1-\pi_0). \tag{5}$$

We define the balk rate p_b of secondary user packets as the average number of secondary user packets being blocked by the system per unit time. The balk rate p_b of secondary user packets can be given as follows:

$$p_b = \sum_{i=0}^{c} \pi_i \sum_{k=c-i+1}^{\infty} \Lambda\alpha(1-\alpha)^{k-1}(k-(c-i)). \tag{6}$$

We define the average delay $E[S]$ of secondary user packets as the average waiting time of the secondary user packets being at the buffer. By using Little's law [8], the average delay $E[S]$ of secondary user packets is given as follows:

$$E[S] = \frac{\beta}{\frac{\Lambda}{\alpha} - p_b} \sum_{i=0}^{c-1} i\pi_{i+1}. \tag{7}$$

4   Numerical Results

In numerical results, we set the whole capacity c of the system to be $c = 10$ as an example. By comparing the numerical results between the batch size parameter $\alpha = 0.3$ and $\alpha = 1.0$, we evaluate the system performance for batch arrival mode and single arrival mode.

By setting the arrival rate of the primary user packets as $\lambda_2 = 0.4$ and the transmission rate of secondary user packets as $\mu = 0.8$, we show how the loss rate p_l of secondary user packets changes with respect to the batch arrival rate Λ of secondary user packets in Fig. 1.

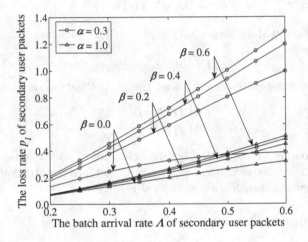

Fig. 1. Loss rate p_l versus batch arrival rate Λ.

In Fig. 1 we observe that for the same batch arrival rate Λ of secondary user packets, the loss rate p_l of secondary user packets will increase as the impatience parameter β of secondary user packets increases. We also observe that for the same impatience parameter β of secondary user packets, the loss rate p_l of secondary user packets will increase as the batch arrival rate Λ of secondary user packets increases.

By setting the arrival rate of the primary user packets as $\lambda_2 = 0.4$ and the transmission rate of secondary user packets as $\mu = 0.1$, we show how the balk rate p_b of the secondary user packets changes with the batch arrival rate Λ of secondary user packets in Fig. 2.

In Fig. 2 we find that for the same batch arrival rate Λ of secondary user packets, the balk rate p_b of secondary user packets will decrease as the impatience parameter β of secondary user packets increases. We also find that for the same impatience parameter β of secondary user packets, the balk rate p_b of secondary user packets will increase as the batch arrival rate Λ of secondary user packets increases.

By setting the arrival rate of the primary user packets as $\lambda_2 = 0.4$ and the transmission rate of secondary user packets $\mu = 0.8$, we show how the average

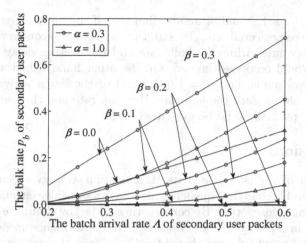

Fig. 2. Balk rate p_b versus batch arrival rate Λ.

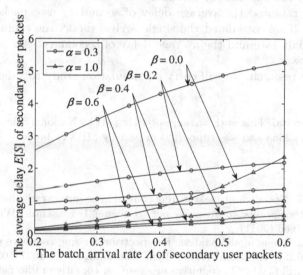

Fig. 3. The average delay $E[S]$ versus batch arrival rate Λ.

delay $E[S]$ of secondary user packets changes with the batch arrival rate Λ of secondary user packets for different impatience parameter β in Fig. 3.

In Fig. 3 we observe that for the same batch arrival rate Λ of secondary user packets, the average delay $E[S]$ of secondary user packets will decrease as the impatience parameter β of secondary user packets increases. We also see that for the same impatience parameter β of secondary user packets, the average delay $E[S]$ of secondary user packets will increase as the batch arrival rate Λ of secondary user packets increases.

Concluding Figs. 1, 2 and 3, we also find that if the impatience of secondary user packets were neglected, i.e., $\beta = 0.0$, the loss rate of secondary user packets would be undervalued, while the balk rate and the average delay of secondary user packets would be overestimated. On the other hand, when we considered the single arrival mode, i.e., $\alpha = 1.0$, instead of the batch arrival mode, i.e., $\alpha < 1.0$, the evaluation for the loss rate, the balk rate and the average delay of secondary user packets would be evaluated lower.

5   Conclusions

For the purpose of evaluating the dynamic allocation strategy in practical CRNs, we established a batch arrival queueing model with possible reneging and potential transmission interruption. By constructing a Markov chain, we analyzed the system model in steady state and obtained the performance measures accordingly. Numerical results showed that if the impatience of secondary user packets were neglected, the loss rate of secondary user packets would be undervalued, while the balk rate and the average delay of secondary user packets would be overestimated. If we considered the single arrival mode, the evaluation for the loss rate, the balk rate and the average delay of secondary user packets would be evaluated lower.

As a future research, we will carry on simulation study to validate the analytical model.

Acknowledgments. This work was supported in part by National Science Foundation (No. 61472342), China and was supported in part by MEXT, Japan.

References

1. Li, H., Han, Z.: Socially optimal queuing control in cognitive radio networks subject to service interruptions: to queue or not to queue? IEEE Trans. Wirel. Commun. **10**(5), 1656–1666 (2011)
2. Abdallah, K.: Queue model analysis for spectrum sharing cognitive systems under outage probability constraint. Wirel. Pers. Commun. **73**(3), 1021–1035 (2013)
3. Oklander, B., Sidi, M.: On cognitive processes in cognitive radio networks. Wirel. Netw. **20**(2), 319–330 (2014)
4. Cadeau, W., Li, X., Xiong, C.: Markov model based jamming and anti-jamming performance analysis for cognitive radio networks. Commun. Netw. **6**(2), 76–85 (2014)
5. Wang, J., Huang, A., Wang, W., Quek, T.: Admission control in cognitive radio networks with finite queue and user impatience. IEEE Wirel. Commun. Lett. **2**(2), 175–178 (2013)
6. Wang, J., Zhang, F.: Equilibrium analysis of the observable queues with balking and delayed repairs. Appl. Math. Comput. **218**(6), 2716–2729 (2011)
7. Hassin, R., Haviv, M.: To Queue or Not To Queue: Equilibrium Behavior in Queueing Systems. Springer, Boston (2003)
8. Jin, S., Yue, W., Saffer, S.: Analysis and optimization of a gated polling based spectrum allocation mechanism in cognitive radio networks. J. Ind. Manag. Optim. **12**(2), 687–702 (2015)

Ensembled Support Vector Machines
for Meta-Modeling

Yeboon Yun[1]([⊠]) and Hirotaka Nakayama[2]

[1] Kansai University, Osaka, Japan
yeboon@kansai-u.ac.jp
[2] Konan University, Kobe, Japan
nakayama@konan-u.ac.jp

Abstract. In many practical engineering problems, function forms cannot be given explicitly in terms of decision variables, but the value of functions can be evaluated for given decision variables through some experiments such as structural analysis, fluid mechanic analysis and so on. These experiments are usually expensive. In such cases, therefore, meta-models are usually constructed on the basis of a less number of samples. Those meta-models are improved in sequence by adding a few samples at one time in order to obtain a good approximate model with as a less number of samples as possible. Support vector machines (SVM) can be effectively applied to meta-modeling. In practical implementation of SVMs, however, it is important to tune parameters of kernels appropriately. Usually, cross validation (CV) techniques are applied to this purpose. However, CV techniques require a lot of function evaluation, which is not realistic in many real engineering problems. This paper shows that applying ensembled support vector machines makes it possible to tune parameters automatically within reasonable computation time.

Keywords: Support vector machines · Bagging · Boosting · Gauss function · Parameter tuning

1 Introduction

In many optimization problems with unknown functions whose evaluation is expensive, it is difficult to get a precise mathematical model. In this case, generally we use a meta-model (or surrogate model) constructed based on a less number of samples. Those meta-models are improved in sequence by adding a few samples at one time (it is called a sequential approximate optimization) in order to obtain a good approximate model with as a less number of samples as possible. It has been observed that computational intelligence can be effectively applied to meta-modeling [8]. Above all, SVM (support vector machine) is promising for this purpose. However, the performance of SVM depends on values of parameters such as the width of Gaussian function. Therefore, it is important to choose appropriate values for the width of Gaussian kernels. For this purpose, one of most popular methods for estimating parameters is cross validation (CV) test. However, CV test is usually time consuming.

© Springer Nature Singapore Pte Ltd. 2016
J. Chen et al. (Eds.): KSS 2016, CCIS 660, pp. 203–212, 2016.
DOI: 10.1007/978-981-10-2857-1_18

On the other hand, several simple formulas [3, 4, 6] have been suggested to estimate the width of Gaussian function roughly. Moreover, it has been shown that ensemble learning such as boosting and bagging has the effect on reducing the sensitivity to parameters [7].

In this research, we propose a sequential learning method using bagging and boosting in order to reduce the burden on choosing the width of Gaussian function. In addition, through numerical experiments, the effectiveness of the proposed method is investigated in terms of generalization ability and control of parameter.

2 Support Vector Regression

Here, we introduce support vector regression (SVR) in brief. Originally, support vector machine (SVM) was proposed by Vapnik *et al.* [1] in the middle of 90's, and has been recognized as one of the most effective tool for machine learning [2, 10]. Later, SVM was extended to SVR for a regression problem introducing ε-insensitive loss function [12] in order to avoid over-learning.

Denote a training data set by $(x_i, y_i), i = 1, \ldots, \ell$, where y_i is an output value for an input vector x_i which is transformed to z_i by some nonlinear mapping.

Various SVR models have been suggested. In this paper, we use the following μ-SVR [9]: for given parameters $\mu > 0$ and $\varepsilon \geq 0$,

$$
\begin{aligned}
&\underset{w,b,\xi,\xi'}{\text{minimize}} && \frac{1}{2} w^T w + \mu(\xi + \xi') \\
&\text{subject to} && (w^T z_i + b) - y_i \leq \varepsilon + \xi, \ i = 1, \ldots, \ell, \\
& && y_i - (w^T z_i + b) \leq \varepsilon + \xi', \ i = 1, \ldots, \ell, \\
& && \xi, \ \xi' \geq 0,
\end{aligned}
\tag{1}
$$

where μ is a parameter to control the amount of the errors ξ and ξ'.

$$
\begin{aligned}
&\underset{\alpha'_i, \alpha_i}{\text{maximize}} && -\frac{1}{2} \sum_{i,j=1}^{\ell} (\alpha'_i - \alpha_i)(\alpha'_j - \alpha_j) z_i^T z_j + \sum_{i=1}^{\ell} (\alpha'_j - \alpha_i) y_i - \varepsilon \sum_{i=1}^{\ell} (\alpha'_j + \alpha_i) \\
&\text{subject to} && \sum_{i=1}^{\ell} (\alpha'_i - \alpha_i) = 0, \\
& && \sum_{i=1}^{\ell} \alpha'_j \leq \mu, \ \sum_{i=1}^{\ell} \alpha_i \leq \mu, \\
& && \alpha'_i, \ \alpha_i \geq 0, \ i = 1, \ldots, \ell.
\end{aligned}
\tag{2}
$$

The inner product term in the objective function of the dual problem (2) can be calculated by using some kernel functions, for example Gaussian kernel function [9]:

$$z_i^T z_j \equiv K(x_i, x_j) = \exp\left(-\frac{\|x_i - x_j\|^2}{2r^2}\right) \tag{3}$$

Then, an approximate function by SVR can be expressed by

$$\hat{f}(x) = \sum_{i=1}^{\ell} (\alpha_i' - \alpha_i) \exp\left(-\frac{\|x - x_i\|^2}{2r^2}\right) + b \tag{4}$$

and finding optimal α_i', α_i and b is learning of SVR.

It should be noted that SVM is usually formulated as a quadratic programming problem (QP) and the calculation time of QP increases exponentially as the size of data set increases.

Generally, it is difficult to estimate appropriate parameters, for example the width r in Eq. (3)), which is closely related to generalization ability. In order to reduce the burden on choosing the width of Gauss function as well as the computation time for learning, we proposed several ensemble learning methods [6, 8]. In this research, we utilize both bagging and boosting for regression problems.

3 Ensemble Learning

In this section, we describe the proposed learning method of which the scheme is shown in Fig. 1:

Step 0. Set a training data set

$$S^{(0)} = \{(x_i, y_i)|i = 1, \ldots, \ell\}$$

and initialize weights

$$d_i^{(0)} = \frac{1}{\ell}, i = 1, \ldots, \ell.$$

The width of Gauss kernel function r_0 is given by a relatively large value.

Step 1. At the initial layer, a training data set is decomposed into several subsets with $\bigcup_{k-1}^{M} S_k^{(0)} \subseteq S^{(0)}$ in which the size of $S_k^{(0)}$ is given by n

Then each weak learner $\hat{f}_k^{(0)}\left(x|S_k^{(0)}\right), k = 1, \ldots, M$, is generated by SVR, and the learner of this layer is defined by the mean of weak learners

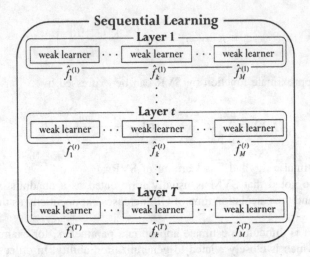

Fig. 1. Scheme of the proposed learning method

$$\hat{f}^{(0)}(x) = \frac{1}{M} \sum_{k=1}^{M} \hat{f}_k^{(0)} \left(x | \mathrm{S}_k^{(0)} \right). \tag{5}$$

Step 2. Set $t := t+1$ and

    2-1 Update the width of Gaussian kernel function which is taken as $r^{(t)} < r^{(t-1)}$. In this research, the following relation will be used:

$$r^{(t)} = 0.5 r^{(t-1)}. \tag{6}$$

    2-2 Update weights $d_i^{(t)}, i = 1, \ldots, \ell$:

$$d_i^{(t)} = \frac{d_i^{(t-1)} \exp\left(-a_{(t-1)} \mathrm{sign}\left(\varepsilon - \left|y_i - \hat{f}^{(t-1)}(x_i)\right|\right)\right)}{\sum_{i=1}^{\ell} d_i^{(t-1)} \exp\left(-a_{(t-1)} \mathrm{sign}\left(\varepsilon - \left|y_i - \hat{f}^{(t-1)}(x_i)\right|\right)\right)}, i = 1, \ldots, \ell, \tag{7}$$

where $\in_{(t-1)} = \sum_{i=1}^{\ell} d_i^{(t-1)} I\left[\left|y_i - \hat{f}^{(t-1)}(x_i)\right| > \varepsilon\right]$ and $a_{(t-1)} = \frac{1}{T} \log\left(\frac{1 - \in_{(t-1)}}{\in_{(t-1)}}\right)$.

    2-3 Update a training data set $S^{(t)}$ and it is decomposed into several subsets $S_k^{(t)}, k = 1, \ldots, M$, by the same way described in Step 1:

$$S^{(t)} = \left\{ (x_i, y_i - \hat{f}^{(t-1)}(x_i)) | i = 1, \ldots, \ell \right\} \tag{8}$$

2-4 Estimate each weak learner $\hat{f}_k^{(t)}\left(x|S_k^{(t)}\right), k = 1, \ldots, M$, by maximizing the revised $\mu-$ SVR formulated by the following:

$$\underset{\alpha_i', \alpha_i}{\text{maximize}} \quad -\frac{1}{2}\sum_{i,j=1}^{\ell} (\alpha_i' - \alpha_i)(\alpha_j' - \alpha_j)K(x_i, x_j)$$

$$+ \sum_{i=1}^{\ell} (\alpha_i' - \alpha_i)y_i - \varepsilon \sum_{i=1}^{\ell} (\alpha_i' + \alpha_i)$$

$$\text{subject to} \quad \sum_{i=1}^{\ell} (\alpha_i' - \alpha_i) = 0, \tag{9}$$

$$\sum_{i=1}^{\ell} \alpha_i' \leq \mu, \; \sum_{i=1}^{\ell} \alpha_i \leq \mu,$$

$$0 \leq \alpha_i' \leq \mu d_i^{(t)}, 0 \leq \alpha_i \leq \mu d_i^{(t)}, \; i = 1, \ldots, \ell.$$

As shown in Eq. (7), the samples x_i with $\left|y_i - \hat{f}^{(t-1)}(x_i)\right| > \varepsilon$ did not be approximated correctly, and thus such those samples gain a larger weight. At the next subsequent layer, it is needed to select a learner that can approximate better for the samples with large weights.

For this purpose, we revised μ-SVR with the constraints that α_i', α_i is limited by the weight $d_i^{(t)}$. This is a parameter to control the amount of influence of the error from the viewpoint of the primal problem to the problem (9). The revised μ-SVR can form a learner paying attention to not well-approximated samples training samples with large weights because those samples at the preceding layer gain larger weights.

2-5 Find the learner of t-th layer:

$$\hat{f}^{(t)}(x) = \hat{f}^{(t-1)}(x) + \frac{1}{M}\sum_{k=1}^{M} \hat{f}_k^{(t)}\left(x|S_k^{(t)}\right) \tag{10}$$

In general, it is known that boosting (the case of the number of weak learner $M = 1$) is susceptible to noisy data and outliers, and sometimes, tends to make over-learning. Combining boosting (the first term in Eq. (10)) and bagging (the second term in Eq. (10)), it is expected that the proposed method can overcome for shortcomings and take advantages of two methods.

Step 3. Terminate if a stop condition is satisfied. Otherwise, go to Step 2.

The convergence of root mean squared error (RMSE) for training data can be used as one of stop conditions. For example, iteration is stopped if the difference between the RMSE of the previous layer and the RMSE of the current layer is less than 10^{-4}.

4   Experiments

Now, the performance of the proposed method is investigated with Sobol's g-function [11], which has been widely used in sensitivity analysis and model surrogating analysis because of its complexity:

$$f(\boldsymbol{x}) = \prod_{i=1}^{s} \frac{|4x_i - 2| + a_i}{1 + a_i},\tag{11}$$

where $x_i \in [0, 1]$ is a design variable and $a_i = \frac{i-2}{2}$ for all $i = 1, \ldots, s$.

For example, Fig. 2 illustrates the function for $s = 2$.

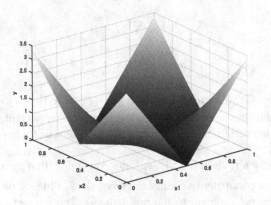

(a) Three dimensional representation

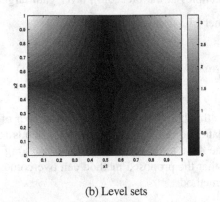

(b) Level sets

Fig. 2. Two dimensional Sobol's g-function

In this experiment, the parameters are used as follows:

- the number of training data: $\ell = 100$, the size of $S_k^{(t)}$: $n = 50$
- the number of weak learners at each layer: $M = 3$
- the initial width of Gaussian function: $r_0 = 2^6$
- $\mu = 10^6, \varepsilon = 0.1$

Training data set of 100 points is generated by an optimized Latin hypercube; test data set is composed of 10^4 points, which are latticed over the hypercube $[0, 1]^2$. Subsets at each layer are generated randomly from the whole data. Meta-models by the proposed method and a single SVR are implemented on MATLAB (optimization tool box - interior point method - is used in solving QP). The performance (generalization ability) of the meta-models is evaluated by root mean squared error (RMSE) for the test data set. The iteration is terminated when the difference between the RMSEs at t and $t - 1$ is less than 10^{-5}.

Figure 3 represents the RMSEs of a single machine over several values of the width r of Gauss kernel function. As can be easily seen from Fig. 3(a), there are appropriate values of r in a very limited area (around $r = 1.0$) and furthermore, the performance (the value of RMSE) is sensitively changing in the neighborhood of $r = 1.0$. This means that it is very difficult to find such an "optimal" value of the width r within a restricted calculation time in real situations because the function evaluation based on experiments is expensive.

In contrast, although the best RMSE of a single machine is slightly better than the one of the proposed method, the proposed ensembled method generates a relatively well approximated function as compared with the meta-model of a single machine. Figure 4(a) represents the RMSE at each layer by the proposed ensemble method, which converges to 0.125 after the layer $t = 9$. Besides, the proposed method provides a stable performance even if the number of weak learners and the initial width of Gaussian function are changed. For example, we obtained the RMSEs between 0.1242 and 0.1247 for three cases with the number of weak learners[1] = 3, 5, 6. We have widely observed in several practical problems that the proposed method provides stable RMSEs both for training and test data, not depending sensitively on the number of weak learners. The learning time of the proposed method seems to increase as the number of weak learners increases. However, the calculation time of QP increases exponentially as the size of data set increases. Since weak learners use training data subsets with a smaller size than the original whole data set, the total computational time of the proposed method has been observed to be less than single machines applying cross validation (CV) test. For example, Table 1 describes the relation of the number of training data versus the calculation time for Sobol's g-function with $s = 2$ (QP is solved by interior point method on MatLab). In order to decide an appropriate value of

---

[1] Depending on the number of data, we should decide the number of weak learners (M) at each layer. However, it is known from our experiences that M should be taken between 3 and 6, and results by the proposed method are not so sensitive to the number of weak learners.

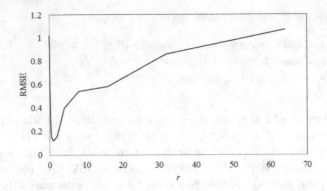

(a) Root mean squared error over various values of r

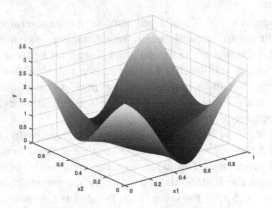

(b) Approximated function ($r = 1.0$)

Fig. 3. The results by a single machine

r, it takes about 56 min (= 22.5 s × 10 times × 15 cases of the width r) by CV test 10 times for 1000 training data. It should be noted, however, in many practical problems that it is very difficult for the single machine to find a proper r (i.e., around 1.0) with only 15 sampled values of r. On the other hand, the proposed method requires around 10 to 11 min under the conditions in which the number of weak learners is 3 and the size of subset data is 750. We can observe that the proposed method makes it possible to reduce the calculation time as well as the burden to find a proper r. We have also investigated several problems, for example, 20 dimensional Sobol's g-function[2] and Frnak's function [5]. For those problems, we have obtained similar results as well.

---

[2] Especially, because Sobol's g-function is strongly anisotropic, it is difficult to provide good performance by using single machine of SVRs [5].

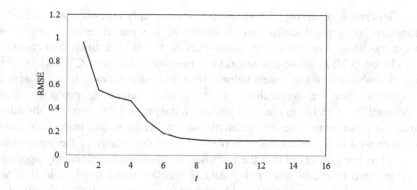

(a) Root mean squared error at each layer

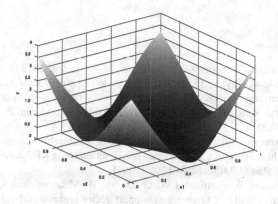

(b) Approximated function

Fig. 4. Results by the proposed ensemble learning method

Table 1. The calculation time to solve QP vs. the number of training data

# data	2000	1000	750	500	100
calculation time	250.0 s	22.5 s	14.0 s	6.5 s	0.6 s

5 Conclusion

In this research, we proposed an effective learning method using both bagging and boosting for regression problems in order to reduce the burden on choosing parameters of kernels for SVMs and to construct meta-models with less computation time. We showed through experiments that the proposed method performs well without paying any particular attention to the value of the width r of Gaussian kernel in SVM in advance.

Generalization ability of SVR models substantially relies on the width of Gaussian function, so a good estimation or choice of the parameter is a critical issue for meta-modeling. Conventionally, parameters of SVMs are estimated or chosen by cross validation (CV) or leave-one-out (LOO) methods. However, CV and LOO require a considerable amount of computational time when the training data is large, and consequently, they are not realistic in many real engineering problems. Considering applicability of SVMs to practical problems, the proposed method has the advantage to make it possible to tune the parameter automatically within reasonable computation time as well as provide a stable performance on the change of the parameters.

It has been observed that the proposed method can be effectively applied to more complicated problems with noisy data (or outliers) and large scale data which are encountered in many practical problems (see in part, e.g., [6]). The details of this topic will be presented in another opportunity.

References

1. Cortes, C., Vapnik, V.: Support vector networks. Mach. Learn. **20**, 273–297 (1995)
2. Cristianini, N., Shawe-Taylor, J.: An Introduction to Support Vector Machines and Other Kernel-based Learning Methods. Cambridge University Press, New York (2000)
3. Haykin, S.: Neural Networks: A Comprehensive Foundation, 2nd edn. Prentice Hall, Upper Saddle River (1998)
4. Kitayama, S., Arakawa, M., Yamazaki, K.: Global optimization by generalized random tunneling algorithm (5th report, approximate optimization using RBF network). Trans. Japan Soc. Mech. Eng. Part C **73**(5), 1299–1306 (2007)
5. Moustapha, M, Sudret, B., Bourinet, J.M., Guillaume, B.: Metamodeling for crashworthiness design: comparative study of kriging and support vector regression. In: Proceedings of the 2nd International Symposium on Uncertainty Quantification and Stochastic Modeling (2014)
6. Nakayama, H., Arakawa, M., Sasaki, R.: Simulation-based optimization using computational intelligence. Optim. Eng. **3**, 201–214 (2002)
7. Nakayama, H., Yun, Y.B., Uno, Y.: Parameter tuning of large scale support vector machines using ensemble learning with applications to imbalanced data sets. IEEE Syst. Man Cybern. Conf. **2012**, 2815–2820 (2012)
8. Nakayama, H., Yun, Y.B., Uno, Y.: Combining predetermined models and SVM/RBFN for regression problems. In: The 6th China-Japan-Korea Joint Symposium on Optimization of Structural and Mechanical Systems, PaperNo. J-64, 8 pages in CD-ROM (2010)
9. Nakayama, H., Yun, Y.B., Yoon, M.: Sequential Approximate Multiobjective Optimization using Computational Intelligence. Springer, Heidelberg (2009)
10. Schölkopf, B., Smola, A.J.: Learning with Kernels: Support Vector Machines, Regularization, Optimization, and Beyond. MIT Press, Cambridge (2002)
11. Surjanovic, S., Bingham, D. http://www.sfu.ca/~ssurjano/gfunc.html. Last Updated: January 2015
12. Vapnik, V.: The Nature of Statistical Learning Theory. Springer, New York (1995)

Deep Context Identification of Deceptive Reviews Using Word Vectors

Wen Zhang[1(✉)], Yipan Jiang[1], and Taketoshi Yoshida[2]

[1] Research Center on Big Data Sciences, Beijing University
of Chemical Technology, Beijing 100029, People's Republic of China
{zhangwen,yipan_jiang}@mail.buct.edu.cn
[2] School of Knowledge Science, Japan Advanced Institute of Science
and Technology, 1-1 Ashahidai, Nomi, Ishikawa 923-1292, Japan
yoshida@jaist.ac.jp

Abstract. This paper proposes deep context by word vectors for deceptive review identification. The basic idea is that since deceptive reviews and truthful reviews are composed by writers without and with real experience, respectively, there should be different contexts of words used by them. Unlike previous work using the whole text collection to learn the word vectors, we produce two numerical vectors for each word by embedding contexts of words in deceptive and truthful reviews separately. Specifically, we propose a representation method called DCWord (Deep Context representation by Word vectors) to use average word vectors derived from deceptive and truthful contexts, respectively, to represent reviews for further classification. Then, we investigate three classifiers as support vector machine (SVM), simple logistic regression (LR) and back propagation neural network (BPNN) to identify the deceptive reviews. Experimental results on the Spam dataset demonstrate that by using the DCWord representation, SVM and LR have produced comparable performance and they outperform BPNN in deceptive review identification. The outcome of this study provides potential implications for online business intelligence in identifying deceptive reviews.

Keywords: Online business intelligence · Skip-gram model · DCWord representation · Deceptive review identification · Deep learning

1 Introduction

With the prevalence of Web 2.0 and social networking, it is widely accepted that for whatever commercial products, the users' opinions are indispensible and valuable for its success of winning good reputation in the market [1, 2]. Online reviews, which refer to users' opinions on a given product which they have using experience in or have something to talk about, are massively emerging in the Internet. These reviews are used by potential customers in purchasing decision making or e-commerce merchants in online promotion campaign. Due to word-of-mouth effect, positive online reviews are helpful for good reputation of products and merchants while negative online reviews will damage its reputation.

On the one side, some merchants endeavor to produce and collect positive online reviews for themselves meanwhile defame their competitors with negative online

© Springer Nature Singapore Pte Ltd. 2016
J. Chen et al. (Eds.): KSS 2016, CCIS 660, pp. 213–224, 2016.
DOI: 10.1007/978-981-10-2857-1_19

reviews, even by hiring "water army" to post online reviews [3]. On the other side, it is impossible to identify deceptive reviews and truthful reviews by human beings satisfactorily [4]. Even worse, anyone can post online reviews anonymously in the Internet with a little cost but may cause great commercial goodness for themselves or great loss for their competitors. This brings about marvelous spam, misleading and even fraudulent online reviews on products and merchants.

With its delivery convenience and price transparency of e-commerce, more and more customers are turning to online shopping [5] in for the sake of time and money saving. Most of them are making use of online reviews in their purchasing decision making, especially for those novices who have no experience on the products in consideration. As the widespread of deceptive online reviews, there are an increasing number of customers who are anxiously worrying about being cheated even trapped by these fake reviews in online shopping. This outcome leads deceptive review identification becoming a hot topic in research field [4, 6–12].

This paper proposes a method called DCWord representation to transfer user reviews into numerical vectors by deep learning method. The basic idea is that there should be different contexts for words in deceptive reviews and truthful reviews. Thus, by deep learning words with its deceptive and truthful contexts, we embed each word into two numeric vectors as deceptive vector and truthful vector. Further, for each review, we represent it using the derived deceptive vectors and truthful vectors based on word occurrences in its content. That is, a DCWord vector is produced for each review by averaging the deceptive and truthful vectors of occurring words. Finally, the DCWord vectors are used to train the classifiers as SVM, LR and BPNN to identify the deceptive reviews. To the best of our knowledge, this paper is the first to introduce word vectors in deep learning to deceptive review identification.

The remaining of the paper is organized as follows. Section 2 presents the word vector representation. Section 3 proposes the DCWord representation methods. Section 4 conducts the experiments. Section 5 concludes the paper.

2   Word Vector Representation

Using numerical vectors to represent words in corpus is becoming an attractive theme in machine learning and NLP areas [13]. Different from traditional methods such as TF-IDF and LSI, where each word can be regarded as with a representation vector of a fixed length of the documents in the corpus, word vectors are produced by learning from its contextual words using neural network architecture and its lengths are variable according to specific application. More importantly, unlike traditional methods that more often than not suffer from the curse of dimensionality, the word vectors contain condensed continuous numbers as elements with much smaller length than the number of documents in the corpus. In this way, the words in texts are embedded in distributed numerical vectors and similar words are represented with similar contexts mathematically.

Usually, the word vectors can be learned by using either continuous Bag-of-Words (CBOW) model or continuous Skip-gram model [14]. For simplicity and computation efficiency, the Skip-gram model is the state of art model adopted by many tasks in natural language processing. For brevity, the Skip-gram model can be depicted in

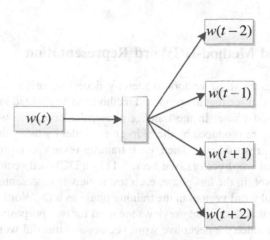

Fig. 1. The Skip-gram model

Fig. 1. Here, we use the word $w(t)$ to predict its 5 maximum distance words $(w(t-2), w(t-1), w(t+1), w(t+2))$. Note that $w(t-2), \ldots, w(t+2)$ are all continuous numerical vectors and projection can be regarded as a neural work with one layer of hidden units.

The goal of the model training is to tune the word vectors that can be used to predict the surrounding words in a sentence. That is, given a sequence of training words $w_1, w_2, \ldots, w_T$, the Skip-gram needs to maximize the average log probability

$$\frac{1}{T} \sum_{t=1}^{T} \sum_{-c \leq j \leq c} \log p(\omega_{t+j} \mid \omega_t). \tag{1}$$

Here, c is the maximum distance from the word ω_t. The larger is c, the more training examples are involved in the model leading to a higher accuracy of the model but with more computation complexity. The basic Skip-gram formulation defines $p(\omega_{t+j} \mid \omega_t)$ using the softmax function

$$p(\omega_{t+j} \mid \omega_t) = \frac{\exp(v'(\omega_{t+j})^T v(\omega_t))}{\sum\limits_{w=1}^{W} \exp(v'(\omega_w)^T v(\omega_t))}. \tag{2}$$

$v'(\omega_w)$ is the output word vector for the word ω_w and W is the size of the vocabulary. In traditional Skip-gram model, there are two word vectors as input word vector $v(\omega_w)$ and output word vector $v'(\omega_w)$ for the word ω_w. In practice, it is very difficult to maximize Eq. 1 if W is huge because, at each run of gradient descent, it will involve updating $O(W)$ parameters. Thus, negative sampling [15] is employed to solve this problem to compute $p(\omega_{t+j} \mid \omega_t)$ within a scope of k native samples. Usually, k is in the range 5–20 are useful for small training datasets, while for large datasets the k can be as small as 2–5.

3   The Proposed Method-DCWord Representation

The framework of using word vectors to identify deceptive reviews automatically is shown in Fig. 2. We can see that the proposed method can be divided into two phases as training phase and test phase. In the training phase, the deceptive word vectors and truthful word vectors are produced by the Skip-gram model by using deceptive reviews and truthful reviews, respectively. Then, each training review is represented by these deceptive and truthful word vectors (see Sect. 3.1) as a DCWord vector to training the model of LR or BPNN. In the test phase, each test review is represented by the learned deceptive and truthful word vectors in the training phase as a DCWord vector to test the learned model. Note that the training reviews are used for two purposes in the proposed method. One is to produce the deceptive word vectors and truthful word vectors by the Skip-gram model and the other is to train the classification model as either LR or BPNN. The test reviews are used for merely to examine the learned classification model.

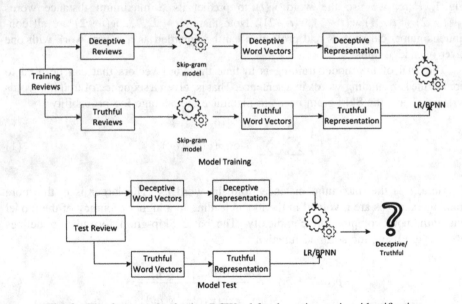

Fig. 2. The framework of using DCWord for deceptive review identification

3.1   DCWord – Text Representation Using Word Vectors

As the input of Skip-gram model is sequences of words, we partition each review as a short passage into sentences using the sentence boundary determination algorithm proposed by [16]. Moreover, we adopt the stop word list obtained from USPTO (United

States Patent and Trademark Office) patent full-text and image database[1] in training the Skip-gram model. For each word w_i in the vocabulary, we use the word sets $window(w_i, c)$ within the window size c (i.e. the maximum distance) in each sentence as the input for the Skip-gram model to produce its neural word embedding as a vector $vec(w)$. Using this method, we derive the deceptive word vectors $vec^{dec}(w)$ and the truthful word vectors $vec^{tru}(w)$ from deceptive reviews and truthful reviews, respectively.

As indicated in Sect. 3, the words in reviews can be regarded as two types. The one is with general concepts such as "room" and "price". The other one is with specific concepts such as "furniture" and "service". For words with general concepts, they are of less effectiveness in discriminating deceptive and truthful reviews because, in both deceptive and truthful reviews, they have very similar contextual information as shown in Table 2. Thus, we need to extract those general words and eliminate them from review representation.

Input:

Dwv -deceptive word vectors;

Twv -truthful word vectors;

N -predefined number of investigated neighbors given a target word;

θ -threshold value for finding general words;

Output:

G - the set of general words in reviews;

Procedure:

1. $CG = \mid Dwv \cap Twv \mid$;

2.   For each word w in CG

3.   Retrieve the top N neighbors of w from Dwv as $dwList$;

4.   Retrieve the top N neighbors of w from Twv as $twList$;

5.   Compute $ol(dwList, twList) = \dfrac{\mid dwList \cap twList \mid}{N}$;

6.   If $ol(dwList, twList) > \theta$

7.   Add w to G;

8.   End if

9.   End for

Algorithm 1. The procedure to extract general words from the reviews.

Algorithm 1 shows the procedure to extract general words from the reviews. Line 1 is used to retrieve the common words from the vocabulary of deceptive reviews and the vocabulary of truthful reviews. Lines 3 and 4 are used to extract its top N deceptive

---

[1] USPTO stop words, online: https://www.uspto.gov/patft//help/stopword.htm.

neighbors *dwList* and truthful neighbors *twList*, respectively, when given a common word w. Line 5 is used to compute the overlapping ratio $ol(dwList, twList)$ of the top N neighbors from deceptive reviews and truthful reviews. With Lines 6 to 8, if the overlapping is greater than the predefined threshold value θ, then the word w should be regarded as a general word. In another word, if a word w was regarded as a general word, it should appear in both deceptive and truthful reviews and have similar contextual words in both types of reviews.

Input:

Dwv -deceptive word vectors;

Twv -truthful word vectors;

$W(r_i)$ -the word list of review r_i;

G - the set of general words in reviews;

Output:

$vec(r_i)$ -DCWord representation for review r_i;

Procedure:

1. Initialize $vec(r_i)$ as a zero vector;

2. For each word w in $W(r_i)$ - G

3.   Retrieve the word vector $vec^{dec}(w)$ in Dwv;

4.   Retrieve the word vector $vec^{tru}(w)$ in Twv;

5. $vec(r_i) = vec(r_i) + (vec^{dec}(w), vec^{tru}(w))$;

6. End for

Algorithm 2. The procedure to represent reviews using the DCWord method.

After deriving the deceptive word vectors $vec^{dec}(w)$ and the truthful word vectors $vec^{tru}(w)$ as well as the general word set G, each review r_i is represented as $vec(r_i) = (vec^{dec}(r_i), vec^{tru}(r_i))$ where $vec^{dec}(r_i)$ is the deceptive representation for the review r_i and $vec^{tru}(r_i)$ is the truthful representation for the review r_i. Specifically, the deceptive representation $vec^{dec}(r_i)$ is computed as $vec^{dec}(r_i) = \frac{1}{|W(r_i) - G|} \sum_{w \in W(r_i) - G} vec^{dec}(w)$ where $W(r_i)$ is the words contained in the review r_i. By analogy, we compute the truthful representation as $vec^{tru}(r_i) = \frac{1}{|W(r_i) - G|} \sum_{w \in W(r_i) - G} vec^{tru}(w)$. Note that if a word only appearing in deceptive reviews, then its truthful vector is simply regulated as a zero vector. In a word, when using DCWord representation, each review r_i is represented as $vec(r_i) = (\frac{1}{|W(r_i)-G|} \sum_{w \in W(r_i)-G} vec^{dec}(w), \frac{1}{|W(r_i)-G|} \sum_{w \in W(r_i)-G} vec^{tru}(w))$. Algorithm 2 shows the details of the procedure to represent reviews using the DCWord method.

3.2 Deceptive Review Identification

In constructing the learning models, we denote the label of deceptive reviews as 1 and the label of truth labels as -1. With SVM as the classifier, we use $vec(r_i)$ as the input vector and the output is the label given by the decision function $f(x) = \text{sgn}((\omega, x) + b) \in \{-1, 1\}$. The linear kernel as $(u * v)^1$ is used to train the classification model because it is proved superior to non-linear kernels in text categorization [17]. With LR as the classifier, we use $vec(r_i)$ as the input vector and the output is the probability $P(y_i = 1 \mid vec(r_i))$, i.e., the probability of the review r_i belonging to the "deceptive" category. If $P(y_i = 1 \mid vec(r_i))$ is greater than 0.5, then the review r_i is regarded as "deceptive". Otherwise, the review r_i is regarded as "truthful". With the BPNN as learning model, the size of the input layer is set as equal to the size of the DCWord vectors of reviews. The size of hidden nodes is set as 100 by trial and error. The size of the output layer is set as 2, i.e. a two-element vector to indicate the possibility of being "deceptive" or "truthful". When using the deceptive reviews for training, its output vector is set as (1, -1) and for truthful reviews, its output vector is set as (-1, 1). For a test review r_i, if we find the value of the first element is larger than the second, it will be labeled as "deceptive". Otherwise, the test review r_i will be labeled as "truthful".

4 Experiments

4.1 The Dataset

The dataset used in the experiments is the Spam dataset from Myle Ott et al. [4]. For each review, we conduct part-of speech analysis, stop-word elimination, and stemming and PCFG analysis. The part of speech of English word is determined by QTAG which is a probabilistic parts-of-speech tagger and can be downloaded freely online[2]. We use the same stop-words as used in training the Skip-gram model. The porter stemming algorithm is used to produce individual word stem[3]. For each review, we extract all of its sentences using the sentence boundary determination method described in [16]. Table 1 shows the basic information of reviews in the Spam dataset.

---

[2] QTag for English part-of-speech, online: http://www.english.bham.ac.uk/staff/omason/software/qtag. html.

[3] Porter stemming algorithm, online: http://tartarus.org/martin/PorterStemmer/.

Table 1. The basic information of reviews in the spam dataset.

Polarity		# of Hotels	# of Reviews	# of Sentences
Positive	Deceptive_from_MTurk	20	400	3043
	Truthful_from_Web	20	400	3480
Negative	Deceptive_from_MTurk	20	400	4149
	Truthful_from_Web	20	400	4483

4.2 Experiment Setup

We divide each of the mentioned datasets into 10 folds uniformly using random sampling and set the retaining level from 0.1 to 0.9 with interval as 0.1 for performance comparison. For instance, with the retaining level as 0.1, we randomly sample 1 of the 10 folds, i.e. 10 percentages of the whole dataset, for classifier training and the remaining 9 folds for classifier testing. We repeat the experiments 10 times to gage the performances.

As mentioned in Sect. 3.1, a crucial parameter for the DCWord representation method is the length of the word vectors derived from the Skip-gram model. Here, we set three different numbers as the layer sizes of the Skip-gram model as 50, 100, 200, 300 and 400 to produce different lengths of word vectors. For instance, if the layer size is set as 50, each word in reviews will be denoted as a numeric vector with length as 100 in DCWord representation (i.e., 50 from deceptive reviews and 50 from truthful reviews).

4.3 Results

Table 2 shows the experimental results of deceptive review identification on the Spam dataset. We set different lengths of word vectors (L.V.) as 50, 100, 200, 300 and 400 and different retaining levels (L.V.) from 0.1 to 0.9. The three adopted classifiers are SVM, LR and BPNN. The decimals are the average accuracies with standard deviations under specific settings of retaining level, length of word vector and classifier.

We can see from Table 2 that when the length of word vector is set as 50, the classifiers SVM, LR and BPNN have produced comparable performances to each other in deceptive review identification. However, when the length of word vector is set no less than 100, we see that SVM and LR outperform BPNN significantly (Wilcoxon sign-rank test, $P < 0.05$). We explain that when the length of word vector is small, all words are embedded in a space with a small number of dimensions. In this case, it seems that all words in reviews share similar contexts. This outcome results in that it is hard to discriminate the reviews by averaging vectors of words in them. That is, there is very limited knowledge (or patterns) that can be learned by the classifiers for deceptive review identification. Even if we add more training data by increasing the retaining level, those vectors are mixed with each other and it is of limited effectiveness in increasing the performances of the classifiers. Nevertheless, when we increase the length of word vector, the words are embedded in a large space. In this case, words with similar contexts are projected with neighborhood coordinates while words with dissimilar contexts are projected with distant coordinates. Thus, it is not very hard to identify deceptive reviews and truthful reviews.

Table 2. The accuracies of DCWord representation in deceptive review identification with different classifiers by setting different retaining levels of training dataset and different sizes of word vectors in Skip-gram model.

L.V. = 50									
R.L.	0.1	0.2	0.3	0.4	0.5	0.6	0.7	0.8	0.9
SVM	0.7104	0.7523	0.7768	0.7750	**0.8219**	0.8083	0.7750	0.7687	0.7520
	±0.1276	±0.1397	±0.1235	±0.1071	**±0.1183**	±0.1281	±0.1026	±0.0874	±0.0752
LR	0.7111	0.7500	0.7902	0.8093	**0.8313**	0.8146	0.7958	0.7875	0.7813
	±0.1156	±0.1072	±0.0982	±0.0876	**±0.1012**	±0.1034	±0.0935	±0.0820	±0.1031
BPNN	0.7049	0.7500	0.8018	0.7958	**0.8275**	0. 8191	0.7850	0.7562	0.7406
	±0.1037	±0.0913	±0.1132	±0.1016	**±0.1170**	±0.0873	±0.0924	±0.0851	±0.0776

L.V. = 100									
R.L.	0.1	0.2	0.3	0.4	0.5	0.6	0.7	0.8	0.9
SVM	0.7301	0.7679	0.7977	0.8300	0.8510	**0.8631**	0.8358	0.8159	0.7875
	±0.1263	±0.1047	±0.1123	±0.0951	±0.1095	**±0.0811**	±0.1194	±0.0893	±0.0725
LR	0.7257	0.7570	0.7938	0.8263	0.8450	**0.8660**	0.8231	0.7959	0.7750
	±0.1178	±0.1016	±0.0868	±0.0879	±0.0929	**±0.1106**	±0.0793	±0.1032	±0.0638
BPNN	0.7130	0.7328	0.7798	0.7966	0.8254	**0.8400**	0.8042	0.7813	0.7719
	±0.1018	±0.1108	±0.1201	±0.1134	±0.0947	**±0.1083**	±0.0917	±0.0761	±0.0953

L.V. = 200									
R.L.	0.1	0.2	0.3	0.4	0.5	0.6	0.7	0.8	0.9
SVM	0.7563	0.7773	0.7889	0.8260	**0.8675**	0.8921	0.8503	0.8188	0.7688
	±0.1161	±0.0893	±0.1011	±0.0942	**±0.0858**	±0.0912	±0.1010	±0.0863	±0.0952
LR	0.7431	0.7791	0.7814	0.8308	0.8860	**0.8883**	0.8511	0.8318	0.7853
	±0.1092	±0.0839	±0.1012	±0.0981	±0.1013	**±0.0911**	±0.1015	±0.1013	±0.0768
BPNN	0.7168	0.7368	0.7595	0.8089	0.8307	**0.8565**	0.8275	0.7864	0.7628
	±0.1204	±0.1018	±0.0831	±0.0927	±0.0738	**±0.0881**	±0.0912	±0.0791	±0.0689

L.V. = 300									
R.L.	0.1	0.2	0.3	0.4	0.5	0.6	0.7	0.8	0.9
SVM	0.7700	0.7972	0.8158	0.8645	**0.9038**	0.9019	0.8751	0. 8247	0.7835
	±0.1180	±0.0813	±0.1037	±0.1037	**±0.1135**	±0.0961	±0.0879	±0.0741	±0.0671
LR	0.7811	0.8200	0. 8437	0. 8778	0.8913	**0.9046**	0.8935	0.8595	0. 7875
	±0.1141	±0.0974	±0.1012	±0.0851	±0.0924	**±0.1013**	±0.0931	±0.0872	±0.0837
BPNN	0.7500	0.7613	0.7886	0.8158	0.8575	**0.8615**	0. 8337	0.8013	0.7545
	±0.1068	±0.1012	±0.0981	±0.0911	±0.1121	**±0.0901**	±0.0836	±0.0971	±0.0791

L.V. = 400									
R.L.	0.1	0.2	0.3	0.4	0.5	0.6	0.7	0.8	0.9
SVM	0.7618	0.7881	0.8418	0.8771	0.9156	**0.9254**	0.8875	0.8334	0.8100
	±0.1044	±0.0938	±0.0872	±0.0936	±0.0763	**±0.0813**	±0.0831	±0.0924	±0.0659
LR	0.7894	0.8302	0.8529	0.8990	0.9263	**0.9288**	0.8988	0.8625	0.8207
	±0.1130	±0.1204	±0.1031	±0.1130	±0.1121	**±0.0873**	±0.0914	±0.1012	±0.0901
BPNN	0.7213	0.7661	0.7800	0.8283	**0.8671**	0.8650	0.8325	0.8094	0.7812
	±0.1112	±0.0918	±0.1012	±0.0924	**±0.0913**	±0.1120	±0.1013	±0.1006	±0.0914

It is very interesting to see that the performances of SVM and LR are comparable to each other in all cases. As reported by Vincent et al. [18] and Hinton et al. [19], LR performs well in prediction task when combined with deep learning models such as denoising autoencoders and restricted Boltzman machine. Thus, it is not surprising that LR performs well in deceptive review identification using word vectors produced by the Skip-gram model. However, it is unexpected that SVM performs comparably to LR because SVM is not usually adopted as a classifier in deep learning. Moreover, BPNN performs worse in comparison with SVM and LR. We explain that when using deep learning techniques to produce the input vectors for classification, LR is capable of this kind of task as a simple classifier. However, this does not mean that SVM, as a more complicated classifier than LR, is not good at the task in this scenario. We conjecture that the inability of BPNN in deceptive review identification because it has similar mechanism with the Skip-gram model by using neural network learning. Thus, the repeated adoption of word embedding by neurons deteriorates the classification performance.

Also, we see from Table 2 that the performances of classifiers in deceptive review identification are maximized at retaining level as 0.5 or 0.6. That is to say, one half of the whole dataset is enough for the classifiers to learning the complete knowledge inherent in the reviews represented by DCWord vectors. After checking the DCWord vectors manually, we find that when the retaining level is small as 0.1 or 0.2, the words in reviews are similar to each other with ambiguities among them. The same outcome is observed when the retaining level is setting large (0.8 or 0.9). We explain that when there is not enough data for training word vectors in the Skip-gram model, the word vectors are very similar to each other due to the correlation sparseness of words. Nevertheless, when giving too much data, the word vectors are also similar to each other due to the abundances of correlation among words. Thus, it is very crucial to tune an appropriate retaining level to learn the DCWord vectors and, we empirically observe that the retaining level should be among 0.4 to 0.6 for deceptive review identification.

We notice that the performances derived in the paper are slightly better than that reported by Feng et al. [9] using n-grams and lexicalized production rules as the feature set for review representation and SVM as the classifier. Moreover, the derived performances are comparable to that of Feng and Hirst [10] by bidirectional alignment on the profile compatibility of the training reviews and the test reviews. We admit that despite the word vectors can capture the contextual information of word to some extent, they cannot characterize the detailed information of the opinion target [20] which is also crucial to identify deceptive and truthful reviews. Thus, how to combine word contextual information and word specific information to further improve deceptive review identification is an ongoing problem.

5   Concluding Remarks

This paper proposes a new representation method called DCWord representation for deceptive review identification by deep learning the word contextual information. With the skip-gram model, we embed each word in the reviews using two numeric vectors: one is from deceptive contextual information and the other is from truthful contextual information. Using the two kinds of numeric vectors, DCWord representation vectors are produced for the reviews. Three classifiers as SVM, LR and BPNN are adopted to identify the deceptive reviews. The experiments on the Spam dataset show that when the lengths of word vectors are no less than 100, SVM and LR outperforms BPNN. The best performance is derived when we set the length of word vector as 400 and the retaining level as 0.5 by the LR classifier. We also explain the experimental results in the paper.

We also find that although the DCWord representation can capture the contextual information of words satisfactorily, it cannot characterize the word specific details precisely. This drawback makes its performances cannot outperform the state of the art technique as profile compatibility in deceptive review identification. In the future, we will further improve the DCWord representation method to embed it with specific details such as term weighting and Entropy measure to enhance its discriminative power in deceptive review identification.

Acknowledgment. This research was supported in part by National Natural Science Foundation of China under Grant Nos. 71101138, 61379046, 91218301, 91318302 and 61432001; Beijing Natural Science Fund under Grant No. 4122087; the Fundamental Research Funds for the Central Universities (buctrc201504).

References

1. Chen, L., Wang, F.: Preference-based clustering reviews for augmenting e-commerce recommendation. Knowl. Based Syst. **50**, 44–59 (2013)
2. Marrese-Taylor, E., Velásquez, J.D., Bravo-Marquez, F., Matsuo, Y.: Identifying customer preferences about tourism products using an aspect-based opinion mining approach. Procedia Comput. Sci. **22**, 182–191 (2013)
3. B. Liu.: Opinion Spam Detection: Detecting Fake Reviews and Reviewers. https://www.cs.uic.edu/~liub/FBS/fake-reviews.html
4. Ott, M., Choi, Y., Cardie, C., Hancock, J.T.: Finding deceptive opinion spam by any stretch of the imagination. In: Proceedings of the 49th Annual Meeting of the Association for Computational Linguistics, Portland, Oregon, pp. 309–319, 19–24 June 2011
5. Lim, Y.J., Osman, A., Salahuddin, S.N., Romle, A.R., Abdullah, S.: Factors influencing online shopping behavior: the mediating role of purchase intention. Procedia Econ. Finan. **35**, 401–410 (2016)
6. Jindal, N., Liu, B.: Opinion spam and analysis. In: Proceedings of WSDM 2008 (2008)
7. Gokhman, S., Hancock, J., Prabhu, P., Ott, M., Cardie, C.: In search of a gold standard in studies of deception. In: Proceedings of the EACL 2012 Workshop on Computational Approaches to Deception Detection, Avignon, France, pp. 23–30, 23–27 April 2012
8. Li, J., Ott, M., Cardie, C., Hovy, E.: Towards a general rule for identifying deceptive opinion spam. In: Proceedings of the 52nd Annual Meeting of the Association for Computational Linguistics, pp. 1566–1576 (2014)
9. Feng, S., Banerjee, R., Choi, Y.: Syntactic stylometry for deception detection. In: Proceedings of the 50th Annual Meeting of the Association for Computational Linguistics, Jeju, Republic of Korea, pp. 171–175, 8–14 July 2012

10. Feng, V.W., Hirst, G.: Detecting deceptive opinions with profile compatibility. In: International Joint Conference on Natural Language Processing, Nagoya, Japan, pp. 338–346, 14–18 October 2013
11. Zhou, L., Shi, Y., Zhang, D.: A statistical language modeling approach to online deception detection. IEEE Trans. Knowl. Data Eng. **20**(8), 1077–1081 (2008)
12. Li, F., Huang, M., Yang, Y., Zhu, X.: Learning to identifying review spam. In: Proceedings of IJCAI 2011 (2011)
13. Collobert, R., Weston, J.: A unified architecture for natural language processing: deep neural networks with multitask learning. In: Proceedings of the 25th International Conference on Machine Learning, pp. 160–167. ACM (2008)
14. Mikolov, T., Sutskever, I., Chen, K., Corrado, G., Dean, J.: Distributed Representations of Words and Phrases and their Compositionality. arXiv:1310.4546 (2013)
15. Mikolov, T., Chen, K., Corrado, G., Dean, J.: Efficient Estimation of Word Representations in Vector Space. arXiv:1301.3781 (2013)
16. Nitin, I., Fred, J.D., Zhang, T.: Text mining: predictive methods for analyzing unstructured information, pp. 15–37. Springer Science and Business Media, Inc., New York (2005)
17. Zhang, W., Yoshida, T., Tang, X.: Text classification based on multi-word with support vector machine. Knowl. Based Syst. **21**(8), 879–886 (2008)
18. Vincent, P., Larochelle, H., Lajoie, I., Bengio, Y., Manzagol, P.: Stacked denoising autoencoders: learning useful representations in a deep network with a local denoising criterion. J. Mach. Learn. Res. **11**, 3371–3408 (2010)
19. Hinton, G.E., Salakhutdinov, R.: Reducing the dimensionality of data with neural networks. Science **313**(5786), 504–507 (2006)
20. Liu, Q., Gao, Z., Liu, B., Zhang, Y.: A logic programming approach to aspect extraction in opinion mining. In: Proceedings of IEEE/WIC/ACM International Conference on Web Intelligence (WI-2013) (2013)

Performance Comparison of TF*IDF, LDA and Paragraph Vector for Document Classification

Jindong Chen[1], Pengjia Yuan[2], Xiaoji Zhou[1], and Xijin Tang[2(✉)]

[1] China Academy of Aerospace Systems Science and Engineering,
Beijing 100048, People's Republic of China
j.chen@amss.ac.cn, zh_xj@sina.com
[2] Institute of Systems Science, Academy of Mathematics and Systems Science,
Chinese Academy of Sciences, Beijing 100190, People's Republic of China
ypj1992@126.com, xjtang@iss.ac.cn

Abstract. To meet the fast and effective requirements of document classification in Web 2.0, the most direct strategy is to reduce the dimension of document representation without much information loss. Topic model and neural network language model are two main strategies to represent document in a low-dimensional space. To compare the effectiveness of bag-of-words, topic model and neural network language model for document classification, TF*IDF, latent Dirichlet allocation (LDA) and Paragraph Vector model are selected. Based on the generated vectors of these three methods, support vector machine classifiers are developed respectively. The performances of these three methods on English and Chinese document collections are evaluated. The experimental results show that TF*IDF outperforms LDA and Paragraph Vector, but the high-dimensional vectors take up much time and memory. Furthermore, through cross validation, the results reveal that stop words elimination and the size of training samples significantly affect the performances of LDA and Paragraph Vector, and Paragraph Vector displays its potential to overwhelm two other methods. Finally, the suggestions related with stop words elimination and data size for LDA and Paragraph Vector training are provided.

Keywords: TF*IDF · LDA · Paragraph vector · Support vector machine · Document classification

1 Introduction

Text is an important source of information, which mainly includes unstructured and semi-structured information. In Web 2.0 era, Internet users are willing to express their opinions online, which accelerates the expansion of text information [1]. Owing to the increasing amount of text information, especially for the unstructured information, to extract useful information or knowledge efficiently, document classification plays an important role [2]. Normally, document classification is to assign the predefined labels to new documents based on the model learned from a trained set of labels and documents.

© Springer Nature Singapore Pte Ltd. 2016
J. Chen et al. (Eds.): KSS 2016, CCIS 660, pp. 225–235, 2016.
DOI: 10.1007/978-981-10-2857-1_20

The process of document classification can be divided into two parts: document representation and classifier training. Compared to classifier training, document representation is the central problem for document classification. Document representation is tried to transfer text information into a machine understandable format without information loss, such as n-gram models. Unfortunately, if n is more than 5, the huge computation cost makes the transformation infeasible. Consequently, several frequently used types of n-gram models are unigram, bigram or trigram [3]. For academic document classification or news classification, owing to the difference of feature words in different categories, those kinds of methods are capable of meeting the requirements of practical application. Meanwhile, a comprehensive analysis of the performances of different classifiers on different data sets is conducted by Manuel et al., and reveals that support vector machine (SVM) and random forests are more effective for most classification tasks [4].

The rapid increase of text data brings new challenges to the available traditional methods [5]. Big corpus dramatically increases the dimension of the representations generated by the traditional methods. High-dimensional vectors take up more memory space, even cannot work on low-configuration computer. Furthermore, even if the transformation is available, the big time cost of classifier training on high-dimensional vectors is another issue for document classification. To meet the tendency of information expansion, it is an important task to reduce the dimension of the representation without much information loss for document classification.

Up to date, document classification is not limited for news classification or academic document classification, and expands to more areas, such as sentiment classification [6], emotion classification [7] and societal risk classification [8]. Different from traditional document classification, these types of document classification face two new challenges: one is that the category of document is related with syntax and word order, the other is different categories may use similar feature words. The traditional methods lack in semantic and word order information extraction, which affects their performances in these areas.

To improve the efficiency of document classification, from dimension reduction and semantic information extraction aspects, several strategies of document representation are proposed:

(1) Topic model. Topic model is not only increasing the efficiency by a more compact topic representation, but also capable of removing noise such as synonymy, polysemy or rare term use. The distinguished methods of topic model include: latent sematic analysis (LSA), probabilistic latent semantic analysis (PLSA) and latent Dirichlet allocation (LDA) [9]. LDA is a generative document model that is capable of dimension reduction as well as topic modeling, and shows better performance than LSA and PLSA. LDA models every topic as a distribution over the words of the vocabulary, and every document as a distribution over the topics, thereby one can use the latent topic mixture of a document as a reduced representation. Based on the representation of latent topic mixture, document clustering and document classification are conducted [10, 11].

(2) Neural network language model. Bengio et al. proposed a distributed vector representation generated by neural network language model [12]. Due to the fixed

and small size of document vector, the distributed representation of neural network language model eliminates the curse of dimensionality problem. Meanwhile, through sliding-window training mode, the semantic and word order information are encoded in the distributed vector space. Recently, based on the neural network language model proposed for word vector construction [13], Le and Mikolov [14] proposed a more sensible method Paragraph Vector (PV) to realize the distributed representation of paragraph or document. Combined with an additional paragraph vector, the method includes two models: PV-DM and PV-DBOW for paragraph or document representation, where the paragraph vector contributes to predict the next word in many contexts sampled from the paragraph.

The purpose of this research is to study the efficiency of different methods for document classification. TF*IDF, LDA and PV have been proposed for a while, and Andrew et al. [15] has compared these three methods on two big datasets: Wiki documents and arXiv articles, each contains nearly 1 million documents, but there is no comprehensive comparative study on these methods for Chinese documents and different sizes of datasets, and no result is reported concerning their classification performances on semantic classification etc. Therefore, to further analyze the performances of these three methods, three datasets: Reuters-21578[1], Sogou news dataset[2] and the posts of Tianya Zatan Board[3] are selected, which includes English and Chinese documents, and aims for news classification and societal risk classification tasks. Based on the document representations generated by these three methods, SVM is adopted for document classification respectively [8], and the performances of each method are compared.

Afterward, LDA relies on the occurrence of words to extract topics, and PV model generates document vector based on word semantic and word order, so stop words present different impacts to LDA and PV. Hence, to clarify the impacts of stop words to LDA and PV, on Sogou news dataset, the influences of stop words elimination operation to LDA and PV model training are analyzed. Next, due to the iterative learning process of PV model, the size of training samples affects the performance of PV. Therefore, on Reuters-21578, Sogou new dataset with repeated data, the performances of PV-SVM are analyzed.

Therefore, the rest of this paper is organized as follows. The data sets and experimental procedures are explained in Sect. 2. The results and discussions are presented in Sect. 3. Finally, concluding remarks are given in Sect. 4.

2 Data Sets and Experimental Procedure

This section introduces data sets and experimental procedures for the different classification algorithms.

---

[1] http://ronaldo.cs.tcd.ie/esslli07/data/reuters21578-xml/.

[2] www.sogou.com/labs/dl/c.html.

[3] http://bbs.tianya.cn/list-free-1.shtml.

2.1   Data Sets

Reuters-21578. Reuters document collection is applied as our experimental data. It appeared as Reuters-22173 in 1991 and was indexed with 135 categories by personnel from Reuters Ltd. in 1996. For convenience, the documents from 4 categories, "agriculture", "crude", "trade" and "interest" are selected. In this study, 626 documents from agriculture, 627 documents from crude, 511 documents from interest and 549 documents from trade are assigned as our target data set.

Sogou. Sogou news dataset used in experiments of this paper are from Sogou Laboratory Corpus. Sogou Laboratory Corpus contains roughly 80,000 news documents, which are equally divided into 10 categories. The categories are Cars, Finance, Education, IT, Healthy, Sport, Recruitment, Culture, Military and Tour.

Tianya Zatan. With the spider system of our group [16], the daily new posts and updated posts are downloaded and parsed. According to the framework of societal risks constructed by socio psychology researchers [17] before Beijing Olympic Games, the new posts of Tianya Zatan in 2012 are almost labeled. To reveal the effectiveness of different methods for societal risk classification of BBS posts, the labeled posts of Dec. 2011–Mar. 2012 are used. The amount of posts of these four months and the amount of posts in different societal risk categories of each month are presented in Table 1. Different from previous two datasets, the figures in Table 1 show the risk distributions of the posts are unbalanced. The posts on Tianya Zatan mainly concentrate on risk free, government management, public morality and daily life, the total number of these categories is more than 85 % of all posts.

Table 1. The risk distribution of posts on Tianya Zatan board of different months

Period / Risk Category	Dec.2011	Jan.2012	Feb.2012	Mar.2012
Risk free	1278	2047	2645	14569
Government Management	3373	1809	3099	6879
Public Morality	3337	3730	8715	6065
Social Stability	954	1013	1746	2108
Daily Life	2641	3063	3142	6920
Resources & Environments	223	147	309	329
Economy & Finance	248	133	460	609
Nation's Security	71	90	214	467
Total	12125	12032	20330	37946

2.2   Experimental Procedures

On the three datasets, three kinds of experiments are tested here: (1) SVM based on TF*IDF method (TF*IDF-SVM), (2) SVM based on LDA method (LDA-SVM), (3) SVM based on Paragraph Vector model (PV-SVM). The desktop computer for all experiments are 64-bit, 3.6 GHz, 8 cores and 16 GB RAM.

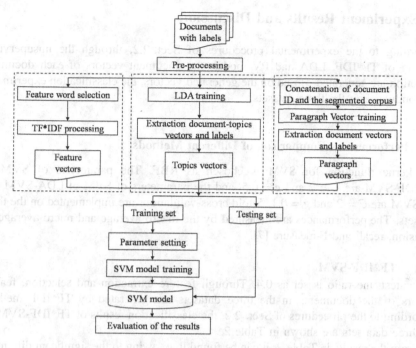

Fig. 1. The process of TF*IDF-SVM, LDA-SVM and PV-SVM for document classification

The pre-processing of English document includes: tokenizing, elimination of stop words and stemming. Meanwhile, the main pre-processing step of Chinese document is word segmentation, and the elimination of stop words is depending on different requirements. Word segmentation tool is Ansj-Seg[4], the stop words dictionary are from Harbin Institute of Technology.

The processes of TF*IDF-SVM, LDA-SVM and PV-SVM for document classification are illustrated in Fig. 1.

The main steps of TF*IDF-SVM include: preprocessing, feature word selection, TF*IDF processing, SVM training and testing and results evaluation. The CHI-square test is adopted for feature word selection. Considering the multi-class classification issue in this field, the One-Against-One approach is adopted.

The main difference of LDA-SVM is the LDA training and topic vectors extraction. The parameters a and β of LDA are set as 1.0/(number of topics). Based on the mixture topic vectors, SVM is also used for document classification. SVM training adopts the same strategy used by TF*IDF-SVM.

For PV-SVM, after the pre-processing of document, an extra document ID is concatenated with the segmented corpus. The processed corpus is fed into PV model to generate the paragraph vector of document. SVM classifier training is based on the generated paragraph vector.

---

[4] Ansj_Seg tool is a JAVA package based on inner kernel of ICTCLAS. https://github.com/ansjsun/ansj_seg.

3 Experiment Results and Discussions

According to the experimental procedures of Sect. 2.2, through the unsupervised training of TF*IDF, LDA and PV model, the document vectors of each document collection are generated. Based on the generated vectors, the classification experiments are conducted on the three datasets.

3.1 Performances Comparison of Different Methods

The kernel function for SVM is chosen as RBF. The parameters of SVM of TF*IDF-SVM are $C = 2$ and $g = 0.5$, and the parameters of SVM of LDA-SVM and PV-SVM are $C = 2$ and $g = 0.1$. 5-fold cross-validations are implemented on the three datasets. The performances are measured by the macro average and micro average on precision, recall and F-measure [7].

3.1.1 TF*IDF-SVM

For χ^2-test, the ratio is set as 0.4. Through feature extraction and selection, feature vectors of the documents in the three datasets are generated by TF*IDF method. According to the procedures of Sect. 2.2, the classification results of TF*IDF-SVM on the three data sets are shown in Table 2.

From the results in Table 2, it can be found that, owing to the significant difference of feature words in different news categories, TF*IDF-SVM shows better performances on news classification. The low-quality corpus of Tianya Zatan and the semantic understanding of societal risk classification decrease the performance of TF*IDF-SVM significantly.

Table 2. The *Macro_F* and *Micro_F* of TF*IDF-SVM

Reuter	1st fold	2nd fold	3rd fold	4th fold	5th fold	Mean
Macro_F	96.23 %	96.36 %	97.87 %	97.62 %	97.30 %	**97.07 %**
Micro_F	95.90 %	96.98 %	96.54 %	95.90 %	96.31 %	**96.32 %**
Sogou	1st fold	2nd fold	3rd fold	4th fold	5th fold	Mean
Macro_F	92.83 %	92.47 %	92.53 %	91.95 %	88.91 %	**91.74 %**
Micro_F	89.49 %	89.46 %	88.73 %	88.71 %	85.99 %	**88.47 %**
Tianya Zatan	1st fold	2nd fold	3rd fold	4th fold	5th fold	Mean
Macro_F	53.89 %	54.85 %	53.66 %	53.45 %	54.84 %	**54.14 %**
Micro_F	60.52 %	60.91 %	60.30 %	60.60 %	61.15 %	**60.69 %**

3.1.2 LDA-SVM

Through the unsupervised training of LDA model, the mixture topic representations of the documents in the datasets are yielded. To reveal the influences of the number of topics, the performances of the numbers of topics: 50, 100, 150, 200, 250, 300 are tested and compared. According to the procedures of Sect. 2.2, the classification results of LDA-SVM on the three data sets are shown in Table 3.

Table 3. The *Macro_F* and *Micro_F* of LDA-SVM

Reuters	The number of Topics	50	100	150	200	250	300
	Macro_F	93.64 %	**94.48 %**	91.52 %	93.61 %	94.41 %	93.52 %
	Micro_F	91.53 %	**92.91 %**	91.92 %	92.78 %	92.35 %	93.00 %
Sogou news datatset	The number of Topics	50	100	150	200	250	300
	Macro_F	75.95 %	80.09 %	85.79 %	86.15 %	85.35 %	**87.23 %**
	Micro_F	75.06 %	79.26 %	80.78 %	81.62 %	81.42 %	**82.11 %**
Tianya Zatan	The number of Topics	50	100	150	200	250	300
	Macro_F	36.56 %	42.26 %	42.19 %	41.17 %	40.15 %	**43.50 %**
	Micro_F	52.51 %	54.33 %	54.30 %	54.65 %	54.44 %	**54.73 %**

From Table 3, it can be found that, with the increase of the number of topics, the improved performances of LDA-SVM are shown on the three datasets. A significant improvement is appeared from 50 to 100, and the differences of other cases become smaller. A similar result is obtained by Andrew [15].

3.1.3 PV-SVM

Through the unsupervised training of PV model, the distributed representations of the documents in the data set are generated. Except for Tianya Zatan dataset, only the labeled documents are used for PV model training. To train PV model on Tianya Zatan dataset, the new posts (title+text) of Dec. 2011–Mar. 2013, more than 470 thousands posts, are used.

To reveal the influences of vector sizes, the performances of the vector sizes: 50, 100, 150, 200, 250 and 300 are tested and compared. According to the procedures of Sect. 2.2, the classification results of PV-SVM on the three data sets are shown in Table 4.

Table 4. The *Macro_F* and *Micro_F* of PV-SVM

Reuters	Vector size	50	100	150	200	250	300
	Macro_F	85.06 %	88.14 %	88.30 %	**88.75 %**	88.66 %	88.42 %
	Micro_F	85.52 %	88.02 %	88.46 %	**88.59 %**	88.54 %	88.41 %
Sogou news datatset	Vector size	50	100	150	200	250	300
	Macro_F	63.10 %	70.16 %	75.29 %	79.25 %	83.34 %	**86.16 %**
	Micro_F	61.27 %	68.09 %	72.13 %	75.35 %	78.17 %	**80.40 %**
Tianya Zatan	Vector size	50	100	150	200	250	300
	Macro_F	35.77 %	44.79 %	46.25 %	47.26 %	47.86 %	**48.20 %**
	Micro_F	53.48 %	55.16 %	55.85 %	56.36 %	56.74 %	**57.03 %**

From Table 4, it can be found that, on the three datasets, with the increase of vector size, the performances of PV-SVM are improved. However, the improvements of *Macro_F* and *Micro_F* are declined, but the improvement tendencies are different for different data sets.

From the results of Tables 2, 3, 4, TF*IDF-SVM obtains overall best performance. Toward Reuters-21578 and Sogou news dataset, the performances of LDA-SVM are better than PV-SVM. Although LDA and PV extract semantic information from documents, the reduced dimension of the two representations loses much information, which leads to the decrease of general performance of LDA-SVM and PV-SVM. However, the dimension of BOW is at least 10 thousands, and the computation and time cost of TF*IDF -SVM are much bigger than the two other methods. Meanwhile, the parameters of SVM are also important to document classification, while this study does not consider the parameter optimization, and the parameters are set by experiences.

3.2   The Influence of Stop Words to LDA and PV

To test the influence of stop words elimination to LDA and PV, Sogou news dataset is selected. Two kinds of experiments are required: (I) the training corpus with stop words; (II) the training corpus without stop words. As the results presented in Sect. 3.1, the performances of LDA and PV model training without stop words have been compared.

In this section, only the experiments of model training with stop words are conducted. To fully compare the performance of LDA-SVM and PV-SVM, two more cases: the number of topics or vector size of 400 and 500 are implemented. The results are shown in Table 5.

Table 5. The *Macro_F* and *Micro_F* of LDA-SVM and PV-SVM for Sogou with Stop Words

LDA-SVM	The number of topics	50	100	150	200	250	300	400	500
	Macro_F	73.82 %	77.26 %	79.50 %	80.46 %	84.42 %	85.02 %	83.50 %	85.71 %
	Micro_F	73.32 %	76.27 %	78.20 %	78.57 %	79.15 %	79.69 %	79.56 %	80.33 %
PV-SVM	Vector size	50	100	150	200	250	300	400	500
	Macro_F	71.72 %	77.28 %	80.73 %	83.80 %	86.36 %	88.18 %	91.28 %	**92.71 %**
	Micro_F	66.69 %	73.06 %	76.95 %	79.96 %	82.33 %	84.23 %	87.42 %	**89.79 %**

As it can be found in Table 5, without stop words elimination, the performance of PV-SVM is more effective than LDA-SVM on Sogou news dataset. However, the results presented in Sect. 3.1, the performance of LDA-SVM is more effective than PV-SVM on Sogou news dataset with stop words elimination. Considering the performances of LDA-SVM and PV-SVM on Sogou with/without stop words, PV-SVM on Sogou news with stop words shows dominant superiority. Meanwhile, the performance of PV-SVM on 500-dimension is also better than TF*IDF-SVM, so PV-SVM may generate better performance than LDA-SVM or TF*IDF-SVM with the increase of dimension.

For LDA, if keeping all stop words, these stop words show similar possibility to all topics, which will decline the clarity of each topic, and affect the performance of LDA-SVM. For PV model, the paragraph token acts as a memory that remembers what is missing from the current context – or the topic of the paragraph. The contexts are fixed-length and sampled from a sliding window over the paragraph for PV model training. In this mode, stop words bring useful information to different documents, and improve the performance of PV-SVM. Therefore, for LDA model training, stop words elimination of the training is necessary, but for PV model training, keeping all words will be more effective.

3.3   The Influence of Data Size to PV

From the previous results, it can be found that LDA model performs better on small datasets: Reuter and Sogou, and PV-SVM obtains better performance on the big dataset: Tianya Zatan dataset, due to almost 50 thousands posts for training. For this reason, to reveal the influence of data size to PV training, the documents of Reuter and Sogou are repeated one and two times for PV training, the results are shown in Table 6 and Table 7.

From Tables 6 and 7, on repeated Reuters-21578 dataset, compared with the non-repeated dataset, the *Macro_F* and *Micro_F* of PV-SVM are significantly increased. A tiny growth of performance is shown from the dataset repeated once to the dataset repeated twice. Conversely, a decrease of *Macro_F* and *Micro_F* on Sogou news dataset is shown, and the more the data repeated, the bigger decrease of performance is generated. As can be found, the data sizes of Reuters-21578 dataset and Sogou news dataset are different, and the size of Reuters-21578 is much smaller than Sogou news dataset. It can be concluded that the training process of PV on Reuters-21578 dataset is under-fitting, so the repeated dataset improves the performance of classification. While the training samples of Sogou news dataset is enough for PV model training, so the repeated data will lead over-fitting to PV model, which only makes worse results. Hence, a proper size of training samples is important to the performance of PV model.

Table 6. The *Macro_F* and *Micro_F* of PV-SVM for Reuters

Reuters repeated once	Vector size	50	100	150	200	250	300
	Macro_F	92.06 %	92.12 %	92.45 %	92.87 %	**93.04 %**	92.29 %
	Micro_F	91.57 %	91.96 %	92.69 %	92.69 %	**93.00 %**	92.65 %
Reuters repeated twice	Vector size	50	100	150	200	250	300
	Macro_F	92.92 %	92.65 %	93.04 %	93.60 %	**93.20 %**	93.19 %
	Micro_F	92.78 %	92.65 %	93.17 %	93.47 %	**93.56 %**	93.52 %

Table 7. The *Macro_F* and *Micro_F* of PV-SVM for Sogou

Sogou repeated once	Vector size	50	100	150	200	250	300
	Macro_F	65.08 %	70.71 %	75.49 %	79.39 %	82.96 %	**85.22 %**
	Micro_F	62.83 %	67.97 %	71.44 %	74.25 %	76.47 %	**78.41 %**
Sogou repeated twice	Vector size	50	100	150	200	250	300
	Macro_F	65.01 %	70.71 %	75.30 %	78.72 %	82.39 %	**84.67 %**
	Micro_F	62.65 %	67.65 %	70.84 %	73.50 %	75.62 %	**77.61 %**

4 Conclusions

In this paper, experiments are conducted to examine the performances of three document representation methods: TF*IDF, LDA and PV for document classification. Basically, two kinds of metrics should be considered: speed and accuracy. Hence, the contributions of this paper can be summarized as follows.

(1) According to the performance comparison of these three strategies on Reuters21578, Sogou news and Tianya Zatan datasets, TF*IDF-SVM shows overall best performance, and LDA-SVM generates better results on small datasets than PV-SVM;
(2) The stop words elimination shows different effects to the performances of LDA-SVM and PV-SVM, and PV-SVM generates much better results when keeping all words, even better than TF*IDF-SVM;
(3) Through the experiments on the repeated training data, it is seen that a proper size of training samples is also important to PV model.

Although we have obtained some preliminary conclusions of TF*IDF, LDA and PV methods, more experiments are required for a comprehensive study. Furthermore, based on the conclusions of this research, how to improve the performance of document classification based on these methods is the future task of this research.

Acknowledgements. This research is supported by National Natural Science Foundation of China under Grant Nos. 61473284, 61379046 and 71371107. The authors would like to thank other members who contribute their effort to the experiments.

References

1. Cao, L.N., Tang, X.J.: Topics and threads of the online public concerns based on Tianya forum. J. Syst. Sci. Syst. Eng. **23**(2), 212–230 (2014). doi:10.1007/s11518-014-5243-z
2. Korde, V., Mahender, C.N.: Text classification and classifiers: a survey. Int. J. Artif. Intel. Appl. **3**(2), 85–99 (2012). doi:10.5121/ijaia.2012.3208
3. Sebastiani, F.: Machine learning in automated text categorization. ACM Comput. Surv. **34** (1), 1–47 (2002). doi:10.1145/505282.505283

4. Manuel, F.D., Eva, C., Senén, B., Dinani, A.: Do we need hundreds of classifiers to solve real world classification problems? J. Mach. Learn. Res. **15**(1), 3133–3181 (2014)
5. Zhang, W., Yoshida, T., Tang, X.J.: A comparative study of TF*IDF, LSI and Multi-words for text classification. Expert Syst. Appl. **38**(3), 2758–2765 (2011). doi:10.1016/j.eswa.2010.08.066
6. Socher, R., Perelygin, A., Wu, J.Y., Chuang, J., Manning, C.D., Ng, A.Y., Potts, C.: Recursive deep models for semantic compositionality over a sentiment treebank. In: Proceedings of the Conference on Empirical Methods in Natural Language Processing *(EMNLP)*, pp. 1631–1642. ACL (2013)
7. Wen, S.Y., Wan, X.J.: Emotion classification in Microblog texts using class sequential rules. In: Proceedings of the Twenty-Eighth AAAI Conference on Artificial Intelligence (Québec, Canada), pp. 187–193. AAAI (2014)
8. Tang, X.J.: Exploring on-line societal risk perception for harmonious society measurement. J. Syst. Sci. Syst. Eng. **22**(4), 469–486 (2013). doi:10.1007/s11518-013-5238-1
9. Blei, D., Ng, A., Jordan, M.: Latent Dirichlet allocation. J. Mach. Learn. Res. **3**(5), 993–1022 (2003)
10. Tang, X.B.: Fang XK (2013) Research on Micro-blog topic retrieval model based on the integration of text clustering with LDA. Info. Stud. Theory Appl. **8**, 85–90 (2013). (in Chinese)
11. Li, K.L., Xie, J., Sun, X., Ma, Y.H., Bai, H.: Multi-class text categorization based on LDA and SVM. Procedia Eng. **15**, 1963–1967 (2011). doi:10.1016/j.proeng.2011.08.366
12. Bengio, Y., Ducharme, R., Vincent, P., Jauvin, C.: A neural probabilistic language model. J. Mach. Learn. Res. **3**, 1137–1155 (2003)
13. Mikolov, T., Chen, K., Corrado, G., Dean, J.: Efficient estimation of word representations in vector space. In: Proceeding of International Conference on Learning Representations (ICLR2013, Scottsdale), pp. 1–12 (2013)
14. Le, Q., Mikolov, T.: Distributed representations of sentences and documents. In: Proceedings of the 31st International Conference on Machine Learning (Beijing). JMLR Workshop and Conference Proceedings, pp. 1188–1196 (2014)
15. Andrew, M.D., Christopher, O., Quoc, V.L.: Document embedding with paragraph vectors. arXiv:1507.07998 (2015)
16. Zhao, Y.L., Tang, X.J.: A preliminary research of pattern of users' behavior based on Tianya forum. In: Wang, S.Y. (eds.) The 14th International Symposium on Knowledge and Systems Sciences, Ningbo, pp. 139–145. JAIST Press (2013)
17. Zheng, R., Shi, K., Li, S.: The influence factors and mechanism of societal risk perception. In: Zhou, J. (ed.) Complex 2009. LNICST, vol. 5, pp. 2266–2275. Springer, Heidelberg (2009)

A New Recommendation Framework
for Accommodations Sharing Based
on User Preference

Qiuyan Zhong[✉], Yangguang Wang, and Yueyang Li

Faculty of Management and Economics, Dalian University of Technology,
116023 Dalian, People's Republic of China
zhongqy@dlut.edu.cn

Abstract. The population of accommodations sharing makes it necessary to make personalized recommendations for users. But general methods perform poorly when applied to accommodations sharing due to severer sparse data and user stickiness. In order to conquer these gaps, this study proposes a specific framework with considering the information of user preference through LDA and Naive Bayes Method. These two methods enable to transfer the user preference into quantitative analysis. Furthermore, LFM possesses the function that can forecast the missing parts through learning user-item rating matrices. It generates the first recommendation list based on the quantized characterization of user preference. The second recommendation list can produce by the Naive Bayes Method. Consequently, the final optimal recommendation result of accommodation sharing is gathered from combining two recommendation lists. Finally, experiments based on the Airbnb real dataset demonstrate the promising potential of this study.

Keywords: Accommodations sharing · Recommendation system · Latent Factor Model (LFM) · Latent Dirichlet Allocation (LDA) · User preference analysis

1 Introduction

With the increasingly population of sharing economy, accommodations sharing develops rapidly, such as Airbnb. The problem that accommodation information overloads is severe because of the massive resources of accommodations and multi-dimensional descriptions. Choosing a desirable accommodation becomes a bothering problem for users. Thus, recommending proper accommodations for them to relieve their heavy burden of selecting becomes a necessary work to be disposed.

Data prepared for recommendation can be mainly classified into two categories: structured and unstructured. The fusion matrix decomposition algorithm can be well used to deal with structured data based on collaborative filtering and other traditional recommendation method. Latent Factor Model (LFM) can achieve automatic clustering based on user behavior matrix, which can avoid the problem of the classification size and dimension selection [1]. There are two major applicable directions of unstructured data processing in the recommendation system: topic mining and emotion analysis. Term Frequency-Inverse Document Frequency (TF-IDF) is the main method applied to

© Springer Nature Singapore Pte Ltd. 2016
J. Chen et al. (Eds.): KSS 2016, CCIS 660, pp. 236–252, 2016.
DOI: 10.1007/978-981-10-2857-1_21

recommendation system in earlier times. The method is simple and the result is too absolute [2]. Blei et al. proposed the Latent Dirichlet Allocation(LDA) mapping the corpus into a document-subject-word Bayesian probability model, which made the extraction of latent subjects more accurately [3, 4]. Naïve Bayes Model (NBM) and Support Vector Machine Model (SVM) are two main analytical algorithms which are widely used to analyze sentiment under the circumstance of no explicit user ratings available [5]. Compared with SVM, NBM has a higher efficient performance and a lower threshold, meanwhile it possesses the advantages of both emotional classification and emotional score quantification by disposing reviews [6, 7]. These methods behave pretty well under the conditions of a small testing data whose information is detailed and concreted, but they behaves poorly and suffer the problem of low-efficiency when the testing data are large enough. Under the background of accommodations sharing, recommendation algorithm needs to dispose massive and multi-dimensional data, at the same time each user's behavior data are few and scattered. Data sparse problem makes traditional methods specializing in the processing of unstructured data performs poorly when they are applied to generate personalized recommendation.

The recommendation of accommodations sharing has some general features which are similar to the traditional hotel recommendation, in addition, it has its own significant difference [8], and they are as follows:

(1) With short rental accommodation market developing rapidly, the total quantity of accommodation will be far more than general hotel, which will cause more serious data sparseness.

(2) Different from standard hotel service, accommodations sharing increased users' dimensions of service experience, user needs to communicate and get along with the house owner. So the factors affecting users' choice are not only like price, cleanliness, etc., but also the expected trust to landlord.

(3) The user loyalty about accommodations is raised as a result of the additional trust relationship's building between users and house owners. In addition, there is a risk of trust when users choose the accommodations that they have never rented before, so they prefer to stay at the accommodations they have lived on the consideration of risk aversion principle, therefore, accommodations sharing has stronger user stickiness.

Hence, the traditional hotel recommendation methods are not applicable. This paper proposes a novel hybrid recommendation framework. Firstly, the user's preference model is reformulated to fill sparse matrix, which effectively alleviates the influence caused by data sparseness on LFM model. Then using NBM to weigh the user's emotion and predict users' stickiness, by which the repeated recommendation generated. Finally, the final recommendation list can be obtained based on the combination of LFM recommendation list and NBM recommendation list.

2 The Acquisition and Preprocessing of House-User's Characteristics

The information of shared accommodations is massive and multidimensional, such as the location, price, facilities, traffic convenience, decoration style, type of housing and so on. All of these descriptions of accommodations can be available from corporation's background database.

In this paper, we select the most important descriptive features of accommodations from the structured data according to users' preferences. Furthermore, by mining user's latent information of preference from their reviews as far as possible, we get a more accurate quantized characterization of users' preference, so recommendation method can have a more high-quality input, which is extremely helpful to the improvement of accuracy and efficiency of the recommendation results.

Let R represent the matrix of user's accommodation behavior, referred as user behavior matrix. Here, the implicit rating for the house $r_{u_i}^j$ is the number of times that the user u_i lives in the house h_j. And $r_{u_i}^{j'}$ correspondingly represents the predicted rating of the house h_j that the user u_i gives. It is common that the user behavior matrix is sparse as the accommodation sites often have lots of registered users and homeowners.

$$R = \begin{bmatrix} r_{u_1}^1 & \cdots & r_{u_1}^n \\ \vdots & \ddots & \vdots \\ r_{u_m}^1 & \cdots & r_{u_m}^n \end{bmatrix}$$

Considering the user stickiness can significantly improve the recommendation's results and adoption rates. In general items recommendation, the item is no need to be recommended again. Since accommodations sharing provides a service experience rather than an item, a user is more likely to choose the same accommodation if he enjoys the service he got before, so accommodations can be recommended repeatedly [11]. The mode of accommodations sharing based on social trust makes user stickiness bigger [12]. The user has an evaluation for every accommodation he has lived. Therefore, this paper quantifies the users' reviews as user stickiness values by mean of an improved Naive Bayes Classifier to analyze their reviews. The c_{uj}^i represents the stickiness value of the house h_j that the user u_i gives. The user stickiness matrix C is shown as the following.

$$C = \begin{bmatrix} c_{u1}^1 & \cdots & c_{uk}^1 \\ \vdots & \ddots & \vdots \\ c_{u1}^m & \cdots & c_{uk}^m \end{bmatrix}$$

3 The Proposed Approach

Based on the house-user's characteristics, the hybrid recommendation method shows as following:

Step 1. Acquire characterization of accommodations. Firstly, get the most important features of accommodations which users care most by using LDA to mine the latent subjects with the most probability from users' review corpus, so each accommodation can be expressed by these features. Furthermore, we get the sentimental mark by means of Naïve Bayes Classifiers to quantify the reviewers' feeling hidden in the reviews. The sentimental mark is also considered as a reference point of accommodations' features. According to different combinational type and processing procedure of these features, three diverse user preference models are proposed.

Step 2. Calculate similarity and generate matrix. This step mainly calculates the similarity between different accommodations by different expression of accommodation's features from different user preference model, by which user behavior matrix can get completed.

Step 3. Generate recommendation list by LFM. LFM predicts the missing parts by learning user's behavior matrix, and generates the first recommendation list after the completion of user behavior matrix is accomplished.

Step 4. Generate recommendation list by repeated recommendation. Repeated recommendation based on Naive Bayesian method generates the second recommendation list.

Step 5. Get final recommendation list. In the end, combine above two recommendation lists together to get the final recommendation list.

3.1 Emotion Analysis Based on the Emotional Dictionary and the Naive Bayes Classifier

In the analysis of emotional comments which belongs to short text analysis on accommodation, Naive Bayes method can get a better performance [13, 14]. The primary reason is that the object of evaluation is relative simplified and the emotional words in comments can represent the emotional tendencies [14]. The degree of the positive emotion can be mirrored by the quantitative emotion value in the way of using Naïve Bayes classifier to analyze comments.

The text d can be expressed as $d = (w_1, w_2, \ldots, w_l)$ based on the feature of emotional dictionary WordNet and the weight of word frequency. The number of feature is l. There are two emotional categories: positive c_p and negative c_n, $C = (c_p, c_n)$. The probability that the text $d = (w_1, w_2, \ldots, w_l)$ belongs to the emotional category $c_j (c_j \in C)$ is denoted as:

$$P(c_j|d) = P(c_j) \prod_{i=1}^{n} P(w_i|c_j)^{wt(w_i)} \tag{1}$$

Here, $P(c_j)$ refers to the prior probability of the category c_j. $wt(w_i)$ is the frequency that the word w_i has appeared in the text d. 3000 comments marked in manual categories are extracted in English from the house dataset in London randomly.

$$P(c_j) = \frac{rev(c_j)}{\sum_{i=1}^{n} rev(c_j)} \qquad (2)$$

Let $rev(c_j)$ denote the number of text that belongs to the category c_j. The posterior probability $P(w_i|c_j)$, referring to the probability of word w_i which appears in the category c_j, is transformed by Laplace to avoid $P(w_i, c_j) = 0$.

$$P(w_i|c_j) = \frac{weight(w_i|c_j) + \delta}{\sum_{i=1}^{n} weight(w_i|c_j) + n\delta} \qquad (3)$$

δ is the minimum value and $weight(w_i, c_j)$ is the sum of weight that w_i belongs to the category c_j. We can get the probability of the positive emotion $P(c_p|d)$ and negative emotion $P(c_n|d)$. So the emotional quantitative value of d is:

$$emotion_d = \frac{P(c_p|d)}{P(c_n|d)} = \frac{P(c_p) \prod_{i=1}^{n} P(w_i|c_p)^{wt(w_i)}}{P(c_n) \prod_{i=1}^{n} P(w_i|c_n)^{wt(w_i)}} = \frac{P(c_p)}{P(c_n)} \prod_{i=1}^{n} \left[\frac{P(w_i|c_p)}{P(w_i|c_n)}\right]^{wt(w_i)} \qquad (4)$$

The emotional quantitative value $emotion_D$ of the text set $D = (d_1, d_2, \ldots, d_n)$ is the average of all values.

$$emotion_D = \frac{\sum_{j=1}^{n} emotion_{d_j}}{n} = \frac{1}{n} \sum_{j=1}^{n} \left\{ \frac{P(c_p)}{P(c_n)} \prod_{i=1}^{n} \left[\frac{P(w_i|c_p)}{P(w_i|c_n)}\right]^{wt(w_i)} \right\} \qquad (5)$$

With (4) and (5), the degree of user's preference of each house he has lived can be calculated. Then the degree of the house that users prefer to live in can be calculated. This degree is regarded as the user's trust on the accommodation.

3.2   The Application of the Model

The comments on the accommodation contain the user's preference information and multiple characteristics of the house. Using these unstructured data as much as possible enables the recommendation system to get more information, which can improve the performance of the recommendation. In this paper, LDA theme model is used to extract features. LDA [4] is one of the most representative theme models. In fact, it is a three-level Bayesian model [8]. Analyzing the text set, we can get the topic-word and document-topic distribution.

Determine the Optimal Number of Topics K in the LDA Model
In this paper, the optimal value of K is estimated by perplexity. Perplexity [16] is a standard measure for evaluating the performance of a language model. The value of perplexity indicates the likelihood estimation in the model that generates new text. It is used to measure the ability of the model to predict the new text. The smaller the value of perplexity, the higher the likelihood estimation is. That is the better performance of the model. The formulation is:

$$Perplexity(D) = exp\left\{ - \frac{\sum_{m=1}^{M} \log_D p(w_m)}{\sum_{m=1}^{M} N_m} \right\} \tag{6}$$

N_m represents for the length of the m th text. $p(w_m)$ is the probability that LDA produces documents. Confusion degree decreases gradually with the increase of number of topics. The stronger the generalization ability of the model with lower confusion, the better performance is.

$$p(w_m) = \sum_d \prod_{n=1}^{N} \sum_{j=1}^{T} p(w_j|z_{j=j}) \times p(z_j = j|w_m) \times p(d) \tag{7}$$

The Similarity Between Houses Based on the Calculation of the Comments
The similarity of the two texts can be generated by calculating the corresponding probability distribution of the subject as the distribution of the text theme is a simple mapping of text vector space. Because the topic is the mixture distribution of the word vector, the KL (Kullback-Leibler) distance is used to measure similarity between two probability distributions [17]. The KL distance is:

$$D_{KL}(p, q) = \sum_{j=1}^{T} p_j \ln \frac{p_j}{q_j} \tag{8}$$

As for all j, when $p_j = q_j D_{KL}(p, q) = 0$. The KL distance is asymmetric. That is $D_{KL}(p, q) \neq D_{KL}(q, p)$. Here we use the symmetric distance:

$$D_\lambda(p, q) = \lambda D_{KL}(p, \lambda p + (1 - \lambda)q) + (1 - \lambda)D_{KL}(q, \lambda p + (1 - \lambda)q) \tag{9}$$

Change Formula (9) into JS (Jensen–Shannon) distance in the case of $\lambda = 0.5$ [18]. The interval of JS distance is $[0, 1]$. We measure the similarity between the texts on the basis of JS distance formula.

$$D_{JS}(p, q) = \frac{1}{2}\left[D_{KL}\left(p, \frac{p+q}{2} \right) + D_{KL}\left(q, \frac{p+q}{2} \right) \right] \tag{10}$$

The comments are divided into two categories: positive and negative. The positive comments are mainly about the advantages of houses, such as complete facilities, the clean and tidy rooms. The negative comments provide relative disadvantages. The Naïve Bayesian classifier is used to quantify the emotional comments firstly to reduce the interference of the LDA model processing as LDA is mainly used to calculate the distinction of underlying text theme. Next the relatively low emotional values will be excluded. Then we can get the house set $H = (h_1, h_2, \ldots, h_n)$ and corresponding comments corpus set Corpora $= (corpora_1, corpora_2, \ldots, corpora_n)$ by combing the rest of the comments into a corpus $corpora_i$ due to the corresponding house h_i. Based on the method of LDA comments corpus set for processing, we can get the probability distribution of the underlying theme about the house. The underlying theme can be understood as property characteristics of different dimensions, such as accommodation, clean degree, the enthusiasm of homeowners and so on. The higher the probability distribution value of the house's underlying theme means, the better characteristic of corresponding dimensions of performance. Therefore, the similarity between houses based on the users' comments can be denoted by the JS distance of house-theme. The similarity computation formula about houses H_p and H_q is:

$$Similarity_{pq} = \frac{1}{D_{js}(p,q)} = \frac{2}{\left[D_{KL}\left(p, \frac{p+q}{2}\right) + D_{KL}\left(q, \frac{p+q}{2}\right)\right]} \tag{11}$$

4 Implementation of Recommendation System

4.1 Description of the Model for Users' Accommodation Preference

For accommodations sharing model, a substantial increasing in the number of accommodation leads to more serious data sparseness, so we introduce content-based recommenders, and conduct content filling: expanding the set of accommodation which users actually live in by similar listings, and a new user behavior matrix is obtained after the filling. The key of "CBF" is to extract the features of accommodation which users are concerned about accurately [10], so it is very important to describe the users' preference as it can reduce the disturbance to the minimum when the matrix is filled.

In this part, we firstly present the description and definition of two simple users' preference models, then on the basis of simple preference model of dimension reduction, we incorporate the LDA model and propose a third form of users' preference models, called "Hybrid-LDA Preference Model". Finally, calculate the similarity between listings of accommodation with these 3 models, and get 3 types of corresponding users' behavior matrix (Fig. 1).

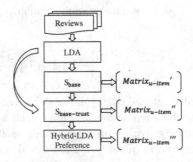

Fig. 1. Matrix completion based on users' preference model

Basic Preference Models

Definition 1. User's preference for accommodation is a multidimensional feature vector $S = (s_1, s_2, s_3 \ldots \ldots s_n)$, s_i is the i th feature that the user concerns about when he chooses the accommodation (ex. price). In order to improve the efficiency of the recommendation system, the paper selects n main features as the user's basic preferences, that is S_{base}. These features can be identified and concluded through the analysis of user comments by LDA.

Based on the analysis of LDA from user reviews, $S_{base} = (s_{price}, s_{rating}, s_{t\_rating})$, $s_{price}, s_{rating}, s_{t\_rating}$ are accommodation price, rating, traffic score respectively. Since the selected accommodation features of this part of the experiments are the basis of the hybrid recommendation method. The results of LDA experiments are shown in this section for the selected accommodation features of the experiments in this part are the basis of the hybrid recommendation method.

In the experiments, the LDA model is used to dispose 1700 English reviews for latent subjective mining, these reviews are extracted randomly from the large corpora. In the meanwhile, the criterion of confusion is used to set the optimum topics K of LDA model, the experiment has shown that the confusion achieved minimum while k = 27, so the optimal quantity of topics is 27. By summarizing the 27 topics manually from the result of LDA, the paper finds that the descriptions of accommodation are focused on price, interior and overall structure, transportation, position and landlord. So in this hybrid recommendation method, four basic features (price, rating, t-rating and position) are selected as the indexes to describe accommodation. The trust of landlord indeed influences the choice of accommodation due to lots of descriptions about landlord in the comments, which also reflect the characteristics of the business model based on trust. Therefore, the Definition 2 is given as following.

The Preference Model Based on Trust

Definition 2. For every accommodation, users' preferences can be divided into basic preferences and trust preferences as follows:

$$S_{base-trust} = (S_{base}, S_{trust}) = (s_{price}, s_{rating}, s_{t\_rating}, S_{trust})$$

Here, s_{trust} is the trust preference of accommodation, which represents the degree of trust. In short accommodation rental mode, the accomplishment of a deal depends on the establishment of a trust relationship, and when choosing an accommodation, the user also needs to consider the expectation of getting along with the landlord. So the higher the s_{trust}, the higher the degree of trust. In addition, s_{trust} is calculated by Naïve Bayesian Method in 3.1, and the specific process is as follows: all corpus collections $Corpora - (corpora_1, corpora_2, \ldots, corpora_n)$ corresponding to set of accommodation $H = (h_1, h_2, \ldots, h_n)$ for accommodation h_i can be gotten by gathering all comments to a corpus collection $corpora_i$. Through the Formula (5) of the Naive Bayesian Method, we can get the quantitative emotional value $emotion_{corpora_i}$, then $s_{trust\_i} = emotion_{corpora_i}$.

Matrix Filling Based on the Simple Preference Model after Two Dimension Reductions

The vectors s_a^h, s_b^h can be used to describe the dimensionless of feature vectors $S_{base}, S_{base-trust}$ corresponding to the two kinds of preference model to avoid the influence on the results caused by orders of magnitude difference. The similarity between accommodation h_a and h_b can calculated by the following formula [19]:

$$sim(h_a, h_b) = \frac{\cos\left(s_a^h, s_b^h\right)}{e^{dis(h_a, h_b)}} \tag{12}$$

Where,

$$dis(h_a, h_b) = \frac{Rad \times \arccos(c) \times \pi}{180}$$

$$c = \sin(lat1) \times \sin(lat2) \times \cos(lon1 - lon2) + \cos(lat1) \times \cos(lat2)$$

$dis(h_a, h_b)$ is the geographical distance between accommodation $h_a(lat1, lon1)$ and accommodation $h_b(lat2, lon2)$, according to the longitude and latitude. This kind of similarity calculation considers both the accommodation features and accommodation position. Because the position is the priority factor in the accommodation selection, which means user will firstly choose the nearest accommodation in intention, so multiplying by $e^{-dis(h_a, h_b)}$ is proper. When the accommodation position matches with the user intention of accommodation position, $e^{-dis(h_a, h_b)}$ equals to 1. The greater the distance deviation of these two position, the smaller $e^{-dis(h_a, h_b)}$. Since $0 < e^{-dis(h_a, h_b)} \leq 1$, the similarity of any two accommodations $sim(h_a, h_b)$ is between 0 and 1. When $sim(h_a, h_b) \geq \partial$ (∂ is the similarity threshold, accommodation h_a is considered similar to h_b. Define that h_a is similar to h_b in the case that $sim(h_a, h_b)$ is no less than the threshold. So we can get the set of similar accommodation H_i' for any h_i of the accommodation set $H(h_1, h_2, \cdots, h_l)$ satisfying the condition that $sim(h_a, h_b)$ is no less than the threshold ∂.

We can get the extended accommodation list, $list_{i\_expand} = H_a' \cup H_b' \cup H_c' \cup \ldots\ldots$, by setting the similarity threshold ∂ to the accommodation lists, $list_i = (h_a, h_b, h_c, \ldots\ldots)$, which is lived by user iso, to the user behavior matrix, $Matrix_{u-item}$, by the extension according to the basic preference S_{base}, we can get $Matrix_{u-item}'$, and by the extension according to the basic preference based on trust $S_{base-trust}$, we can get $Matrix_{u-item}''$.

Hybrid-LDA Preference Model

The user preference for accommodation S is a multidimensional feature vector, involving all aspects of features of accommodation. Generally the user preference for accommodation is described by choosing the most critical of several features for dimension reduction, such as basic preference model S_{base} and preference model based on trust $S_{base-trust}$. This method measures the similarity of accommodation by several critical features with stable performance. But because of ignoring the other features, there exists some errors. Most of the characteristics of accommodation are retained by using LDA to calculate the distribution of the underlying themes (features) of comment to describe preference as the method does not reduce dimension. But the text data-comments and comments coped with LDA have a lot of randomness and volatility, so the performance is unstable. Besides, when the quantity of accommodation is too large, the document of comments will be too large. So, too much noise data will significantly reduce the effects of LDA, and the cost of computation will be very large and the efficiency of recommendation is reduced. Therefore, the effect is not good in calculating the similarity by totally using this method.

To solve these problems, the paper proposes a new preference model by combining simple preference model of dimension reduction with LDA model, so it is named "Hybrid-LDA Preference Model". First of all, the method moderately reduces dimensions for user accommodation preference S, then calculates the similarity between houses using this model. Secondly, it is needed to eliminate most of the objects by setting the appropriate similarity threshold in order to generate the final similarity set of accommodation by using the method of 3.2, which is based on the similarity of comments. In fact, this hybrid model is a method which can reduce the dimension first and then increase the dimension. It greatly reduces the computation of the LDA models, improves the computing efficiency of the recommendation system, and ensures the accuracy and stability of user preferences. Finally, based on this model, we obtain the extended user behavior matrix $Matrix_{u-item}'''$.

4.2   Initial Recommendations Based on LFM Model

After filling the original user behavior matrix, data sparse problem has been largely alleviated. This is a good solution to the problem of the susceptibility of data sparse of LFM models. When the data set is relatively sparse, performance degradation is very obvious, so the extended user behavior matrix can increase recommended performance substantially.

In the training set, we select the accommodation which user has had a record of accommodation as the positive sample and the others without a record of accommodation as the negative sample. In this paper, the number of comments is used as a measure of the hot degree. The negative sample selection principle is that the number is closed to the positive sample. A larger number of accommodation comments are selected as negative sample. According to the LFM model:

$$R = P^T \times Q \tag{13}$$

Where $P \in R^{f \times m}$ and $Q \in R^{f \times n}$ are two matrixes after dimension reduction, here P represents the preference weights of that m users to f hidden features, and Q represents the weights of n accommodations to f hidden features. So the value of rating prediction of user u to accommodation h_i, r_{ui}', can be calculated as following:

$$r'_{ui} = \sum\nolimits_F p_{uf} q_{if} \tag{14}$$

Where $p_{uf} = P(u,f)$, $q_{if} = Q(i,f)$, Then we need to optimize the following loss function to find the most suitable parameters p and q:

$$C(p,q) = \sum_{\substack{u \in U \\ i \in H}} \left(r_{ui} - r'_{ui}\right)^2 = \sum_{\substack{u \in U \\ i \in H}} \left(r_{ui} - \sum_{f=1}^{F} p_{uf} q_{if}\right)^2 + \gamma p_u^2 + \gamma q_i^2 \tag{15}$$

Where $\gamma \|p_u\|^2 + \gamma \|q_i\|^2$ is the regularization item to avoid over fitting. In order to minimize the loss function in the above formula, the Stochastic Gradient Descent method is used here, which is a basic optimization algorithm. The method firstly used to find the fastest decline in the direction of the parameter by partial derivative, then to optimize the parameters by the iterative method continuously [20]. It can recommend the top N in predicted ratings that the user have not been to make the initial recommendation list.

So far, the initial recommendation list is generated through LFM. But as mentioned above, the object of accommodation recommendation is a kind of service, which is different from the item recommendation, so it can be recommended repeatedly. The improved LFM can remove the accommodation that the user lived from the recommendation list. In order to apply it to accommodation recommendation, we need the repeated recommendation considering the features of users.

4.3   Repeated Recommendation Based on User Agglutinant

According to the characteristics of the high user stickiness in the house sharing, we analyze user emotion to make repeated recommendation through the method of Naive Bias Classifier which is proposed in preceding part of the text, so as to improve the precision of recommendation. The set of accommodation which the user u_i has lived is $S_i = \left(s_1^{t_1}, s_2^{t_2}, \ldots s_j^{t_j} \ldots, s_n^{t_n}\right)$, $s_j^{t_j}$ represents that the user lived t_j times in the accommodation s_j times. The comments of u_i are the set $R_i = \left(r_1^{\sum_1^{t_1} reviews}, r_2^{\sum_1^{t_1} reviews}, \ldots r_j^{\sum_1^{t_j} reviews} \ldots, r_n^{\sum_1^{t_n} reviews}\right)$, $r_j^{\sum_1^{t_j} reviews}$ is the set of the t_j comment of u_i to the accommodation s_j. Through Formula (5), get the set of affective ratings $E_i = (emotion_{D_1}, emotion_{D_2}, \ldots, emotion_{D_n})$ of the user u_i to the accommodation S_i, and then pick out the highest M to recommend to the user according to the Top-N principle. There is a great variation among S due to different users, so set the following rules of the creation of repeated recommendation list: to n accommodations the user lived, when $n \leq 3, M = 1; 4 \leq n \leq 6, M = 2; n \geq 7, M = 3$.

4.4 Integration of Two Recommended Lists

Finally, we get the final recommended list by mixing the recommendation list generated by the LFM model and the recommended list generated by repeated recommendation together. The element in the final integrated recommended list is sorted as follows: the elements from repeated recommendation list are in the front, after which are the elements generated from LFM recommended list, all the containing elements are sorted from high to low by the emotional score and predicted score respectively. The final recommendation results are generated based on the principle of Top N.

5 Experiment Result and Analysis

5.1 Dataset and Experiment Environment

The experimental dataset is from the Airbnb of the UK's London, which is available from http://insideairbnb.com/get-the-data.html. The dataset includes detailed records of all transactions including a total of 300,000 records within the period from September 2009 to September 2015. It also gives the descriptive data corresponding to related dimensions. Some columns of this dataset are shown in Table 1.

Table 1. Raw dataset

Id	host_id	latitude	Longitude	Beds	Price	Reviews	Review_scores_location
375799	1889832	51.40	-0.2498	2	$150	......	9
6003865	29230489	51.41	-0.2966	1	$180	......	7

In Table 1, *Id* and *host_id* represent the user and house respectively, *reviews* are the review comments, the original data is non-null, and the last column *review_scores_location* is user's evaluation of convenience of traffic. In order to alleviate the interference caused by too much noise data, data cleaning and noise reduction are conducted, those data whose key information is missing are eliminated. Some special transactions, such as abnormally high prices castle, tree house are also eliminated. The final experimental dataset contains 290,273 transactions generated from 17,694 housing resources and 254,522 users.

At the beginning of the experiment, 1700 English comments with better quality are selected manually. The next operations are about word lowercasing, stemming, stop words & punctuation removing. After these processing, LDA model is used to find the latent active topics from these comments to recognize characteristics of accommodations which users care most about. Based on these important characteristics of houses, the user preference model of dimension reduction is obtained. Furthermore, 3000 English reviews which are randomly sampled from the large dataset are marked positive and negative emotions to train the Naive Bayes model.

In the experimental data of this paper, nearly 90 % of the users only have one transaction. If these data are divided into training set, the test set doesn't include corresponding user behavior. If these data are divided into the test set, a severe cold start problem will arise, and user preference model proposed in this paper can't be able to learn user preferences in this case. In order to alleviate interference, those accommodation records more than twice are selected as the final experimental data, then 60 % of the data are randomly selected as the training set, and the rest as the test set.

The experimental environment is Linux operating system (Ubuntu 15.10), CPU Intel Core i7-4790 (3.2 GHz), 16G RAM, Programming language are R&Python.

5.2 Experiment Result and Analysis

In following experiments, three preference models presented in this paper are tested separately. **Experiment 1**, fill the user behavior preference matrix based on the basic preference (S_{base}); **Experiment 2**, fill the user behavior preference matrix based on the trust preference ($S_{base-trust}$); **Experiment 3**, fill the user behavior preference matrix based on the hybrid preference ($Hybrid - LDAPreferenceModel$).

Result and Analysis of Experiment 1 and 2

According to the two element selectivity of Airbnb user behavior, this paper selects the precision and recall rate as the experimental evaluation index. Experiment 1 and 2 both have the similarity accommodation threshold ∂, here ∂ is set to 0.1, 0.2 … 0.6 separately. The results of Experiment 1-2 are shown in Figs. 2 and 3 separately. There are 6 curves in each of Figs. 2 and 3, each curve denotes the result at different values of accommodation threshold.

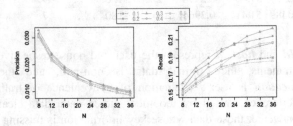

Fig. 2. Precision & recall of experiment 1

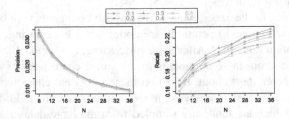

Fig. 3. Precision & recall of experiment 2

By comparing with the two experimental results, the prediction accuracy and recall accuracy both increase as we increase similarity threshold from 0.1 to 0.3, after that they become worse, so the optimum similarity threshold is 0.3. When the similarity threshold is set to 0.1 or 0.2, the user behavior matrix is over-filled, which causes interference to the generation of recommendation. When the similarity threshold is set to 0.4, 0.5 or 0.6, the matrix completion is insufficient. Judging by the trend of these curves in Figs. 2 and 3, we can find that the result will be worse while similarity threshold is set to 1, which means the original user behavior matrix will not be filled, this once again proves the LFM model is sensitive to the sparseness of data. When the similarity thresholds ∂ of Experiment 1 and Experiment 2 are both set to 0.3, the prediction and recall precision of Experiment 2 have increased by nearly 0.036 % ∼ 0.192 % and 0.758 % ∼ 1.076 % respectively comparing with the corresponding performance of Experiment 1. Thus, it proves that the base-trust model can depict user's preference more accurately than base model.

Result and Analysis of Experiment 3

By the comparison of Experiment 1 and 2, $S_{base-trust}$ is selected as the simple preference model of Experiment 3 to carry out for its better performance. By setting the similarity threshold to 0.3, the set of all the potentially similar houses accommodation can be obtained. The LDA model is used to process the corpus of the comments for each potentially similar house accommodation. The experiment has shown that majority of the optimum K of LDA fastens on 30, 35, 40 for the 20 houses selected randomly. So we set threshold value to 30, 35, 40 separately to find the optimum value. The user behavior matrix can be filled by the dataset from the final similar accommodation set composed of the top two-thirds of each accommodation's potential similar houses. In addition, for further comparison, the best recommendation result of Experiment 2, where the threshold is set to 0.3, is also displayed in the Fig. 4, labeled as E-2. The final experimental results are shown in Fig. 4.

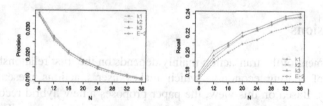

Fig. 4. Precision & recall of experiment 3

There are four curves in each graph of Fig. 4, the k1, k2, k3 denote the results at three different topic value of LDA, the E-2 denotes the best recommendation results of Experiment 2 whose similarity threshold is set to 0.3. We find that Experiment 3 performs better than Experiment 2 no matter what value the theme K is set from k1, k2, or k3. From the experimental results, it can be seen that themes obtained from the recommendation results of k1, k2, k3 is better than that of Experiment 2, no matter the theme K value is k1, k2, or k3. Meanwhile, the results can have the best recommendation effect when the theme K is set to k2, and the recommendation accuracy and recall rate increase about 0.048 % ~ 0.14 % and 0.7 % ~ 1.263 % respectively comparing to the performance of Experiment 2.

The results of Experiment 1-3 show that the recommendation effect of Experiment 3 is better than that of Experiment 2, and the effect of Experiment 2 is better than Experiment 1. Thus, as to all the three available user preference model presented in this paper, the hybrid preference Hybrid-LDA preference model is better than that of trust based preference $S_{base-trust}$, which is better than that of preference S_{base}.

5.3   Comparison with Collaborative Filtering

For further comparison, collaborative filtering method is used to generate the recommendation with the same dataset, the result is shown in the Fig. 5. Collaborative filtering is one of the most fundamental recommendation techniques, which provides personalized recommendations based on users' tastes. In the meanwhile, it suffers from a range of problems, the most fundamental being that of data sparseness. Sparseness in ratings makes the formation of inaccurate neighborhood, thereby resulting in poor recommendations [21].

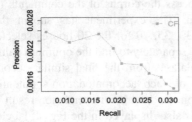

Fig. 5.   Result of CF recommendation

Compared with the poor performance of the collaborative filtering recommendation, the proposed method in this paper remarkably improved the efficiency and the precision. The precision rate has been improved by about 2.8 %, which has shown that accurate matrix completion technique based on the users' preference performs very ideally to address the problem of sparse data sensitivity.

6   Conclusions

The establishment of the transaction mainly depends on the trust relationship under the background of sharing economy, which makes user's actions present some new characteristics. Based on this view, the paper proposes a new hybrid recommendation framework specializing in recommending for accommodations sharing. The framework aims at quantifying users' preference accurately to construct a new user preference model by mining more implicit information of preference from structured data and achieving the most utilization of unstructured data. Furthermore, this paper continue to optimize the user preference model and the hybrid-LDA model can achieve the most ideal result with the most efficiency, with the help of the matrix completion significantly alleviates the problem of the degenerate performance of the recommendation system caused by sparse data. Finally LFM successfully deals with the dilemma of sparse data in recommendation systems, and results in better performance than classic recommendations such as collaborative filtering algorithms.

In this paper, based on an accurate characterization of users' preference, matrix completion successfully breaks through the dilemma of sparse data, so the LFM can get a more ideal result. It proves that an accurate characterization of users' preference is of great significance to the performance of recommendation, utilizing structured data to find users' preference information has a limitation to some extent, more information are hiding in the unstructured data like users' reviews, however, mining latent information from these unstructured data requires a large consumptive calculation. In this paper, the hybrid-LDA model can achieve an ideal result because of its best efficiency, which has a significance of lowering the threshold of using these high-consuming algorithms in actual application. In the future, we will focus on more sophisticated fusion schemes and designing more efficient method to characterize users' preference, so recommendation will be more personalized and will be applied in broader areas.

Acknowledgements. This research is financially supported by NSFC with its projects (71533001).

References

1. Zhu, Y., Shen, X., Ye, C.: Personalized prediction and sparsity pursuit in latent factor models. J. Am. Stat. Assoc. (2015)
2. Ramos, J.: Using TF-IDF to determine word relevance in document queries. In: Proceedings of the First Instructional Conference on Machine Learning (2003)
3. Krestel, R., Fankhauser, P., Nejdl, W.: Latent Dirichlet Allocation for tag recommendation. In: Proceedings of the Third ACM Conference on Recommender Systems, pp. 61–68 (2009)
4. Blei, D.M., Ng, A.Y., Jordan, M.I.: Latent Dirichlet Allocation. J. Mach. Learn. Res. **3**, 993–1022 (2003)
5. Suykens, J.A., Vandewalle, J.: Least squares support vector machine classifiers. Neural Process. Lett. **9**, 293–300 (1999)
6. Pazzani, Michael J., Billsus, Daniel: Content-based recommendation systems. In: Brusilovsky, Peter, Kobsa, Alfred, Nejdl, Wolfgang (eds.) Adaptive Web 2007. LNCS, vol. 4321, pp. 325–341. Springer, Heidelberg (2007)
7. Keerthika, G., Priya, D.S.: Feature subset evaluation and classification using naive Bayes classifier. J. Netw. Commun. Emerg. Technol. (JNCET) **1**(2015). www.jncet.org
8. Yannopoulou, N., Moufahim, M., Bian, X.: User-generated brands and social media: couchsurfing and Airbnb. Contemp. Manage. Res. **9** (2013)
9. Lu, J., Wu, D., Mao, M., Wang, W., Zhang, G.: Recommender system application developments: a survey. Decis. Support Syst. **74**, 12–32 (2015)
10. Zhang, K., Wang, K., Wang, X., Jin, C., Zhou, A.: Hotel recommendation based on user preference analysis. In: 31st IEEE International Conference on Data Engineering Workshops (ICDEW), pp. 134–138 (2015)
11. Garbers, J., Niemann, M., Mochol, M.: A personalized hotel selection engine. In: Proceedings of the Poster Session of 3rd ESWC (2006)
12. Gutt, D., Herrmann,P.: Sharing Means Caring? Hosts' Price Reaction to Rating Visibility (2015). aisel.aisnet.org
13. Go, A., Huang, L., Bhayani, R.: Twitter sentiment analysis. Entropy **17** (2009)

14. Yang, D., Yang, A.M.: Classification approach of Chinese texts sentiment based on semantic lexicon and naive Bayesian. Appl. Res. Comput. 27(10), 3737–3739 & 3743 (2010)
15. Cao, J., Xia, T., Li, J., Zhang, Y., Tang, S.: A density-based method for adaptive LDA model selection. Neurocomputing 72(7–9), 1775–1781 (2009)
16. Cao, J., Zhang, Y.D., Li, J.T., Tang, S.: A method of adaptively selecting best LDA model based on density. Chin. J. Comput. 31(31), 1780–1787 (2008)
17. Wang, Z.Z., Ming, H.E., Du, Y.P.: Text similarity computing based on topic model LDA. Comput. Sci. 40(12), 228–232 (2013)
18. Endres, D.M., Schindelin, J.E.: A new metric for probability distributions. IEEE Trans. Inf. Theory 49(7), 1858–1860 (2003)
19. Liu, S.D., Meng, X.W.: Approach to network services recommendation based on mobile users' location. J. Softw. 25(11), 2556–2574 (2014)
20. Bell, R.M., Koren, Y.Y.: Lessons from the Netflix prize challenge. ACM SIGKDD Explor. Newsl. 9, 75–79 (2007)
21. Liang, C.Y., Leng, Y.J.: Collaborative filtering based on information-theoretic co-clustering. Int. J. Syst. Sci. 45(3), 589–597 (2014)

Author Index

Printed in the United States
By Bookmasters